William Evans

The mammalian fauna of the Edinburgh district

With records of occurrences of the rarer species throughout the

south-east of Scotland generally

William Evans

The mammalian fauna of the Edinburgh district
With records of occurrences of the rarer species throughout the south-east of Scotland generally

ISBN/EAN: 9783337272111

Printed in Europe, USA, Canada, Australia, Japan

Cover: Foto ©berggeist007 / pixelio.de

More available books at **www.hansebooks.com**

THE MAMMALIAN FAUNA

OF THE

EDINBURGH DISTRICT,

WITH

RECORDS OF OCCURRENCES OF THE RARER SPECIES THROUGHOUT
THE SOUTH-EAST OF SCOTLAND GENERALLY

BY

WILLIAM EVANS, F.R.S.E.
SECRETARY TO THE ROYAL PHYSICAL SOCIETY

EDINBURGH: M'FARLANE & ERSKINE
1892

PREFACE

THE present volume is substantially a reprint of a paper forming part of the "Proceedings" of the Royal Physical Society for the Session 1890-91 (vol. xi., pp. 85-171). The paper was read at the meeting of the Society on 15th April 1891; and printed in December, a few additional records and observations being first added to bring it down to date.

The reproduction of the memoir in its present form is largely due to the representations of a number of friends, who have been good enough to say that it supplies a long felt want, and to suggest that, if issued as a separate volume, its usefulness would be materially increased. Having personally experienced the want of a treatise on the Mammalia of the country surrounding the Scottish metropolis, I have the less hesitation in giving effect to their suggestion.

For the permission to reprint the paper, I take this opportunity of expressing my sincere thanks to the Council of the Royal Physical Society. My best thanks are also due to many friends and correspondents—Mr Eagle Clarke in particular—for valuable assistance rendered in a variety of ways. The information I owe to each has been acknowledged, as far as possible, in the text.

W. E.

EDINBURGH, *April* 1892.

THE MAMMALIAN FAUNA

OF THE

EDINBURGH DISTRICT

INTRODUCTORY REMARKS

The late Mr E. R. Alston, in the closing sentences of his Catalogue of the Mammalia of Scotland,[1] pointed out to Scottish naturalists that the distribution of the mammalian-life of the country was much in need of revision, and cited the Shrews, Mice, and Voles as especially deserving of attention. Impressed with the truth of his remarks, I have been endeavouring during the last few years to work out in some degree of detail the distribution in our own neighbourhood of the above-mentioned groups, and also of the Bats, which had been equally neglected. My original intention was to communicate to the Royal Physical Society the results of my observations on these groups alone. Being, however, also in possession of a mass of data bearing on the past and present distribution of the other recent animals of the class Mammalia occurring in the district, I ultimately decided to combine the two sets of notes, and lay them before the Society in the present form.

The "Edinburgh District," as here understood, embraces the valley of the Forth, and such parts of the adjoining areas

[1] Published in 1880 by the Glasgow Natural History Society as part of its "Fauna of Scotland, with special reference to Clydesdale and the western district."

—Tay and Tweed—as lie within a moderate distance, say twenty to thirty miles, of the city, the whole being capable of investigation in the course of a series of easy excursions, seldom requiring more than a day for their accomplishment. It is, in fact, practically the same area as that adopted by Balfour and Sadler in their "Flora of Edinburgh," and shown in the map which accompanies both editions of that work,—a section of country presenting a combination of physical features peculiarly rich and varied. The counties embraced are—on the south, East Lothian, Midlothian, West Lothian, and Peebles, with parts of the adjoining counties of Berwick, Roxburgh, Selkirk, Lanark, and Stirling; and on the north, Fife, Kinross, Clackmannan, and a portion of Perth. Through the centre—from west to east—winds the Forth with its estuary and broad firth, into which innumerable tributary streams from secondary valleys empty their waters. Fresh-water lochs and ponds also abound. The upper part of the main valley, penetrating as it does the south-western section of the Perthshire Grampians, is thoroughly wild and alpine in character. From the rugged mountains of this north-west corner, a series of sub-alpine ranges—the Campsie Fells, the Pentlands, the Moorfoots, and the Lammermoors—with their connecting moorlands, constitute the watershed on the south; while the almost alpine Ochils, the Cleish Hills and the Lomonds, mark it on the north. Between this rampart of hills and the shores of the Forth, every variety of lowland country is to be found—fertile lands and barren commons, green meadows and furze-clad hills, breezy heights and secluded dells, with woods and plantations of deciduous trees and pines on every side. The part of the Tweed area of which we take cognisance lies largely in the pastoral county of Peebles, and consists for the most part of grassy and heather-clad hills, intersected by a multitude of glens dear to the angler. The section of the Tay area falling within our limits is mainly lowland towards the east, and highland in the west. It will thus be seen that the district, whether we contemplate it as in the natural garb of former times, or as now changed in outward aspect by the hand of man, is well fitted to be the home of a mammalian fauna rich both in species and in individuals. In the case of the rarer and more interesting species, occurrences throughout the south-east of Scotland generally will be alluded to.

From the earliest times man has ever exercised a modifying influence on mammalian-faunas, adversely affecting some species either by direct persecution or by rendering the country unsuitable to their habits, and directly or indirectly fostering the increase of others. He has, moreover, long been in the habit of importing certain species from one country or district to another, so that it is not always easy to separate the indigenous from the introduced. The more populous a district becomes, and the more its agricultural industries are developed, the greater will be the changes on its fauna. Add to these factors the existence for many centuries of a large and influential class of landowners holding strong views regarding the preservation of game, and it will readily be understood that the district around Edinburgh was probably the first in Scotland to witness a radical change in the character of its mammalian-life within historic times. The larger predaceous animals, such as the Wolf and the Bear, which carried destruction among the flocks, and even threatened the life of the herdsman himself, would be among the first to succumb. Many species would be hunted for their skins or their flesh; others mainly for sport. The smaller Carnivora would receive further attention on account of their visits to the poultry-yard, and Hares and Rabbits because of injury to the crops. Then came the game laws—another interference with the balance of nature—accelerating the destruction of the predatory animals, and facilitating the increase of the rodents. The inordinate increase of the Rabbit led in its turn to a universal system of trapping to keep it in check, and from that day the fate of most of the remaining terrestrial Carnivora was sealed. Bocce and Sibbald have put us in possession of much valuable information regarding the fauna of Scotland in the sixteenth and seventeenth centuries, but as a rule their statements are too general to be of direct interest in the present inquiry. But from the "Old Statistical Account" of the parishes we get some excellent glimpses into the state of our fauna a century ago. Even then the predatory animals had been in great measure banished to the outlying parishes, where, however, they were still not uncommon, as the following extracts clearly testify:—

CAMPSIE (STIRLINGSHIRE).—After mentioning the Badger and the Fox and their varieties, the writer of the article continues: "There are likewise (on the Campsie Fells) weasels, otters, polecats, hedgehogs, wild cats; and, of late, several martins have been seen among the rocks. . . . It may be observed, that beasts of prey are every day becoming scarcer. Till within these two years, we had a regular bred huntsman who hunted this district; his salary was paid by the tenants, at so much per plough, which huntsman and dogs were kept and fed by each tenant in his turn" (vol. xv., p. 323).

CALLANDER (PERTH).—"Red deer come here for food and shelter in severe winters. Roes breed in our woods. Hares, rabbits, foxes, wild cats, badgers, otters, moles, polecats, weasels, and black martins, are also to be found here" (vol. xi., p. 598).

DOUNE (SOUTH-WEST PERTH).—"The wild animals here

are the same as in the neighbouring parishes, hares, rabbits, foxes, badgers, otters, foumarts, or polecats. The braes on the north-east side of Cambuswallace House have been long a receptacle for badgers and foxes; but these mischievous animals are now much banished" (vol. xx., p. 49).

ALLOA (CLACKMANNAN).—" The wild animals are the same as are common to all the Low Country : hares, rabbits, foxes, badgers, otters, *foumarts*, or polecats, and *stoats*, or ermines. These last are very rare. There are no wild cats" (vol. viii., p. 645).

TILLICOULTRY (CLACKMANNAN).—" The wild quadrupeds are hares, rabbits, foxes, hedge-hogs, weasels, polecats, badgers, and otters" (vol. xv., p. 200).

FOSSAWAY AND TULLIEBOLE (PERTH AND KINROSS).—" Of quadrupeds, there are foxes, badgers, otters, polecats, hares, and rabbits" (vol. xviii., p. 466).

CASTLETOWN (ROXBURGH). — " The wild quadrupeds are foxes, hares, wild cats, polecats, weasels, the white weasel, often seen in winter, hedgehogs, and Norway rats" (xvi., p. 76).

Many of the terrestrial species mentioned in the following pages have doubtless been inhabitants of the district from a remote antiquity. The naturalist, therefore (and I feel sure the sportsman too, if he would but allow himself to look at the matter in all its aspects), cannot contemplate without feelings of regret the extermination of such animals as the Wild Cat, the Marten, the Polecat, and the Badger, whose ancestors were the contemporaries of the Bear, the Wolf, the Wild Boar, and the Beaver, and in all probability inhabited the district while extinct Deer and Oxen, and maybe even the gigantic Mammoth, still lingered on its soil.[1]

---

[1] Evidence of the former existence of the Bear, the Wolf, and the Wild Boar, even within historic times, is not wanting ; and remains of the Wolf have been found on the Pentland Hills above Dreghorn ; of the Beaver at

As the sole object of this memoir is to furnish information concerning the mammals now or recently to be found in the district, I have deemed it expedient to include in the catalogue only those species which are known to have occurred within the present century. Disregarding such as had disappeared before Sibbald's day, and also the Narwhal (*Monodon monoceros*), of which an example, obtained near the Isle of May in June 1648, is mentioned by Tulpius ("Obs. Med.," p. 376), I thus exclude but one species having anything like a substantial claim to a place in the list, namely, the Sperm Whale or Cachelot (*Physeter macrocephalus*), of which three are recorded as having been stranded in the Forth, namely, one at Limekilns in 1689, and two near Cramond, in 1701 and 1769 respectively. The fact that when the Athole herd of the so-called Wild White Cattle was broken up in 1834, about a dozen of them were secured by the then Duke of Buccleuch, and kept for a few years in the park at Dalkeith, does not entitle the animal to a place in this list, even when taken in connection with Bishop Leslie's statement that it existed at Stirling (in the royal park?) in 1578. Those who desire information regarding the extinct forms may be referred to Owen's "British Fossil Mammals and Birds," Dr J. A. Smith's papers in the "Proceedings" of the Society of Antiquaries of Scotland (vols. viii., ix., etc.), Woodward & Sherborn's "Catalogue of

Kimmerghame (Berwickshire), Linton Moss (Roxburghshire), and Loch Marlee (Perthshire); of the Elk at Whitrig Bog (Berwickshire), near Hawick, and near Selkirk; apparently also at Duddingston and near Cramond (Midlothian), near North Berwick (East Lothian), Kirkurd (Peeblesshire), Marlee (Perthshire), etc.; of the Reindeer near Craigton (Linlithgowshire), and on the Pentland Hills above Dreghorn; of the Great Long-horned Ox or Urus at Whitrig Bog and Swinton Mill (Berwickshire), near Jedburgh, Lilliesleaf, and Linton Moss (Roxburghshire), Whitmuirhall near Selkirk, Newburgh (Fife), etc.; and of the Mammoth at Clifton Hall on the confines of the counties of Edinburgh and Linlithgow (Mem. Wern. Soc., iv., 58), and at Kimmerghame (Berwickshire).

British Fossil Vertebrata," and Harting's "British Animals Extinct within Historic Times."

The number of recent species (exclusive of the White Cattle) hitherto recorded on satisfactory evidence for the whole of Scotland is fifty-seven, which includes six additions to Alston's 1880 catalogue, namely, two Bats (*Vespertilio nattereri* and *V. mystacinus*) and four Cetaceans (*Balænoptera borealis, Grampus griseus, Lagenorhynchus albirostris,* and *Delphinus delphis*). Of these, forty-eight have valid claims to a place in the present catalogue, and no fewer than forty-six of them have occurred within the Forth area. This is exclusive of Natterer's Bat, the record of which requires confirmation. The Greenland or Harp Seal (*Phoca grœnlandica*), and the Dormouse (*Muscardinus avellanarius*)— both perhaps less likely to occur—have also been recorded, but on insufficient evidence, and are consequently likewise excluded. The details of the comparison are given in the following table, which also contains the figures for the British Isles as a whole:—

Order.	Britain.	Scotland.	Edinburgh.
Chiroptera,	12 species.[1]	5 species.	3 species.
Insectivora,	5 ,,	5 ,,	5 ,,
Carnivora,	14 ,,	13 ,,	11 ,,
Rodentia,	13 ,,	12 ,,	12 ,,
Ungulata,	3 ,,	3 ,,	3 ,,
Cetacea,	19 ,,	19 ,,	14 ,,
In all,	66 species.	57 species.	48 species.

The number by which the Edinburgh list falls short of the British is largely made up of Bats, while the deficiency, as compared with the Scottish list, consists almost entirely of

---

[1] *V. murinus, V. discolor,* and *V. dasycneme,* being in all likelihood mere importations, are excluded.

Cetaceans and Seals. The British Bats are mainly confined to England, several of them being of rare occurrence even there, so that we can scarcely hope to increase our list by discoveries in this group to the extent of more than two or three. The occurrence of any further marine Carnivora on our coasts is not at all likely; but the identification of one or two additional Cetaceans is probably only a question of time. It is therefore to the Bats and Cetaceans that we must look for any augmentation of the list. The latter group may safely be left in the hands of Sir William Turner, LL.D., F.R.S., whose investigations have already thrown so much light on this section of our fauna. Should any of my readers have opportunities of obtaining Bats from any part of Scotland, they will confer a favour by forwarding specimens either to myself or to Mr Eagle Clarke of the Museum of Science and Art, Edinburgh, for examination.

On account of their timidity and more or less nocturnal habits, comparatively few of our native quadrupeds come under the notice of the casual observer; and the same causes, by rendering the observation and study of them matters of considerable difficulty, are no doubt in great measure responsible for the scanty attention paid to the class by the majority of field naturalists. With the exception of the very meagre lists in Rhind's "Excursions," illustrative of the natural history of the environs of Edinburgh, and in Stark's "Picture of Edinburgh," and of a short article by Mr Eagle Clarke in Pollock's "Dictionary of the Forth," published in May 1891, no account of the Mammalia of the district has hitherto, so far as I am aware, been written. Records and short notices bearing on the subject are, however, by no means scarce, though scattered over a wide field of zoological and other literature. A list of the publications from which information has been derived will be found at the end of the volume.

The arrangement and nomenclature followed are those adopted by Flower and Lydekker in their recently-published "Introduction to the Study of Mammals," with this exception, namely, that I begin with the Chiroptera and end with the Cetacea, instead of the reverse, the result being that the orders are presented in the same sequence as in the second edition of Bell's "History of British Quadrupeds," which is still the standard work on these animals. The following quotations from Flower and Lydekker's book are well worth bearing in mind in this connection. The authors remark, p. 84, that "In systematic descriptions in books, in lists, and catalogues, and in arranging collections, the objects dealt with must be placed in a single linear series. But by no means can such a series be made to coincide with natural affinities. The artificial character of such an arrangement, the constant violation of all true relationships, are the more painfully evident the greater the knowledge of the real structure and affinities. But the necessity is obvious, and all that can be done is to make such an arrangement as little as possible discordant with facts." Again (p. 85), speaking of the sub-class *Eutheria* (*Monodelphia*), with which alone we have here to do, they make the following observations:—"Their affinities with one another are so complex that it is impossible to arrange them serially with any regard to natural affinities. Indeed, each order is now so isolated that it is almost impossible to say what its affinities are; and none of the hitherto proposed associations of the orders into large groups stand the test of critical investigation. All serial arrangements of the orders are therefore perfectly arbitrary; and although it would be of very great convenience for reference in books and museums if some general sequence, such as that here proposed, were generally adopted, such a result can scarcely be expected, since equally good reasons might be given for almost any other combination of the various elements of which the series is composed."

As regards the means adopted for the capture of Mice, Voles, and Shrews, I may mention that I use a small trap known as the "Cyclone mouse-trap"—an American patent—and find it most effective. It consists of a metal plate about two inches square, to which are attached two strong spring "jaws" of single wires, which, when the trap is unset, rest on the edges of the foot-plate, so that the whole instrument occupies very little space, thus permitting of a number being carried in the pocket without inconvenience. When exposed to damp they are apt to become rusty, which impedes the action of the springs, but this is easily obviated by the application of a little oil or vaseline. For bait I have generally used cheese, cake of any sort, or a piece of apple, but on the suggestion of Mr W. D. Roebuck of Leeds, I have lately tried powdered aniseed, and find it remarkably attractive to most of the micro-mammals.

Order CHIROPTERA.

LONG-EARED BAT.

PLECOTUS AURITUS (*L.*).

THOUGH much less abundant than the next species, this Bat is widely distributed in the district, and is by no means rare. I have myself obtained it at Tynefield and Gosford in East Lothian, at Colinton in Midlothian,[1] and at Lamancha in Peeblesshire; and have seen examples (including one obtained by Mr Harvie-Brown at Dunipace in Stirlingshire), or had it reported to me, from a number of other localities on both sides of the Forth. All my captures have been made while the animals were at rest in their hiding-places, which I have invariably found to be about buildings. The Colinton colony inhabit the ruins of an ancient castle, concealing themselves in narrow holes in the masonry of the roofs of the vaults and passages. The species was recorded for Alloa as far back as 1793 ("Old Stat. Acc.," viii., p. 646).

Mr George Pow has sent me a specimen which was taken in the day-time, near Dunbar, on 10th June last. It flew against the face of its captor apparently in a dazed state, and fell helpless to the ground. At Yetholm, in June 1886, I observed one fluttering in broad daylight near an old mill in the same semi-conscious state, and might easily have captured it but for an intervening stream. Its behaviour was in marked contrast to the activity displayed by the Pipistrelle,

---

[1] A specimen obtained at Yester in East Lothian, on 17th October 1891, has been handed to me by Mr Bruce Campbell; and I have to thank Mr D. F. Mackenzie for another taken at Mortonhall, near Edinburgh, on 10th November.

as I have seen it under similar circumstances. As a rule this is a late flier, and hence is seldom noticed on wing.

Since writing the above, I have received from Mr Chouler, gamekeeper, Dalkeith Park, a Long-eared Bat which entered his house on the evening of 1st October (1891), and was observed to take two flies from the ceiling of a room in which a bright light was burning. It is alive still (12th October), and is allowed out of its box for a couple of hours every evening for a flight by gaslight. When first exposed to the light, it seems rather bewildered, but very soon becomes quite lively, flitting about with the utmost confidence, examining every corner of the room, and ever and anon resting suspended head downwards from the cornice or curtains. It delights in scrambling about the pictures, the window-blinds, and even the chairs; and often settles on the floor, where it moves with considerable rapidity (indeed, it may almost be said to run), keeping the body practically clear of the ground. A more knowing little creature I have seldom seen; and, having discovered that there is sufficient space below the room-door for it to creep through, its endeavours to overcome obstacles placed in the way of its escape are most persistent and amusing. Once it flew up the chimney, but in five minutes returned very much begrimed, of course, with soot. At the light it never flies, as moths invariably do. During flight the tail is not, as a rule, stretched full out behind (as represented in most illustrations of bats on wing), but, at about half its length, is curved downwards and forwards. From a flat surface—the table or the floor, for instance—it springs into flight without the slightest difficulty. Flies and small pieces of butcher-meat it takes readily from the hand. The inside of the mouth is quite pallid as compared with that of a Daubenton's Bat I had alive a month ago.[1]

[1] Captured at Cromdale, Strathspey, *vide* Scottish Naturalist, 1891, p. 190.

COMMON BAT or PIPISTRELLE.

Vesperugo pipistrellus (Schreb.).

Bats are familiar objects in almost every part of the district, as they flit to and fro in the twilight of summer evenings; and, of the three kinds positively known to occur, the Pipistrelle is undoubtedly by far the most abundant and generally distributed. I have examined examples from many localities; and, by its size and its style of flight, have identified it on the wing hundreds of times. Some years ago I captured about a dozen in a few minutes with a landing-net at Macbiehill in Peeblesshire, and in May 1890 I secured nearly as many with an insect-net at Gosford in East Lothian. I have also recently netted it at Duddingston Loch. For the opportunity of examining fresh specimens from the following localities, my best thanks are due to the persons whose names are given after each, namely:—Dunbar and East Linton, in Haddingtonshire (G. Pow); Yester, Haddingtonshire (Bruce Campbell); Grant's House, Berwickshire (Bruce Campbell); Stobo, Peeblesshire (J. Thomson); The Inch, Midlothian (T. Speedy); Dalmeny Park, Linlithgowshire (Bruce Campbell); Dunipace, Stirlingshire (J. A. Harvie-Brown); Broomhall, Fifeshire (W. Lumley).

With us the Pipistrelle's usual period of activity is from April till late in October,[1] but it may be seen on wing in mild weather in every month of the year. On New Year's Day 1883, I noticed one flying briskly about Cramond church, and two were seen in Dalmeny Park on 28th January 1891. Occasionally, too, it may be observed abroad at midday. In

---

[1] One was seen at Dalmeny on 7th November 1891, and I saw one flying briskly near Morningside at noon on 23rd December (a sunny and very frosty day), and again in the evening on 28th February and 19th March 1892.

June last I watched one for fully a quarter of an hour flying in the bright sunshine at Broomhall, near Dunfermline, and was much struck with its activity, and the facility with which it evaded stones and other missiles thrown at it. It would appear that it also occasionally travels a considerable distance in search of food, for on 18th September 1884, while waiting for wild-fowl by the sandhills at the mouth of Aberlady Bay, one flew round me several times. Some of the many examples that have passed through my hands have been decidedly paler than the ordinary form, while one or two have been almost black. Their flight, as observed in my room, is more rapid and erratic than that of the last species, and instead of alighting on the cornice or curtains in an inverted position, they settle with the head uppermost, and as a rule only invert themselves when about to take flight again.

DAUBENTON'S, OR THE WATER BAT.

VESPERTILIO DAUBENTONI *Leisl*.

Our knowledge of the distribution of this species in the district is still imperfect, but enough is known to show that it is, at least locally, not uncommon. Under the name of *V. marginatus*, it was recorded from Fife so long ago as 1828 by Fleming ("British Animals," p. 6). During the summer of 1869 I observed a number of Bats flitting over a still reach of the Esk above Penicuik, and one which I succeeded in striking down with a walking-stick proved to be of this species. In the Edinburgh Museum there are three specimens (two adults and a newly-born young one), identified by Mr Eagle Clarke ("Scottish Naturalist," 1891, p. 92), which were taken at The Inch, near Liberton, in July 1880 by Mr T. Speedy, who tells me they were found with many others in the hollow of an old ash. He assures me there were many dozens clustered together in this dormitory, which, however, they forsook immediately after its discovery. I have little doubt the feeding-ground of the colony was Duddingston Loch, over whose surface, on calm summer evenings, numbers of *V. daubentoni* may generally be seen.[1] On two occasions in June of the present year (1891) I was on the south side of the loch at dusk, and identified several pairs, their larger size, and habit of gliding in easy circles close to the surface of the water, serving at once to distinguish them from the Pipistrelles. In other suitable localities I have from time to time seen bats that doubtless were of this species. In the "Proceedings" of the Berwickshire Naturalists' Club (vol. ix., p. 441), Dr Hardy states—on the authority, he tells me, of the late Mr Robert Gray—that Daubenton's Bat is well known about Dunbar, a district from which, through the attention of Mr G. Pow, I have received three examples in the flesh. Two of these—male and female—were shot as they flew over a sheet of water at Broxmouth on 20th June of the present year (1891), and the third by the Tyne at East Linton four days later. The expanse of the wings in the first of these was about $9\frac{3}{4}$ inches, while in the last it was barely 9. The species has likewise been recorded more than once from Roxburghshire ("Proc. Berw. Nat. Club," ix., p. 441 ; and "Scottish Leader" of 22nd June 1888). Its Scottish stronghold, however, would appear to be in the neighbourhood of Dumfries, where Mr Service tells me it is commoner than the Pipistrelle. As in the case of other Bats, July seems to be the usual time for the production of the young.

---

[1] Vogt, in his Natural History of Animals (London, 1887, vol. i., p. 106), says of this Bat, "Its winter quarters are in hollow trees, often pretty far from its hunting-ground."

[NATTERER'S BAT.

Vespertilio nattereri *Kuhl.*

This Bat is recorded as having occurred near Dalkeith, but unfortunately the circumstances are not altogether satisfactory. They are as follows:—On 28th September 1880 the late Mr R. Gray wrote to Mr Harvie-Brown in these terms—"I find a new bat to our Scottish lists in some plenty near Dalkeith, viz., *R.* [sic] *nattereri.*" ... "*Nattereri* was in dozens in the hole of a tree," statements which were published by Mr Harvie-Brown in the "Proceedings" of the Glasgow Natural History Society (vol. iv., p. 303). It seems strange that Mr Gray, who was always so solicitous for the full and proper recording of rarities, should have let the subject drop here if he was convinced of the correctness of his identification. I have endeavoured to follow the matter up, but with little success. No specimen of Bat from Dalkeith, or of *V. nattereri* from any locality, can be found in Mr Gray's collections. From Mr Hope, taxidermist, Edinburgh, I learn, however, that in 1880 he had in his shop, besides those brought by Mr Speedy from The Inch, and since referred by Mr Eagle Clarke to *V. daubentoni*, some Bats from Dalkeith Park which Mr Gray remarked were of an uncommon kind, and one of which was given to him at his request. On inquiry at Mr Malcolm Dunn, Dalkeith Gardens, he informs me that "in the spring of 1886 a large colony of bats (roughly estimated at about fifty) were discovered one sunny afternoon thickly clustered beneath the 'roan' and eave, and behind a rain-pipe" in a corner of his house. He mentioned the circumstance to Mr Gray, who expressed the opinion that they would belong to the species known as Natterer's Bat. For two or three years they frequented the same corner, but have now entirely deserted it; and though Mr Dunn has made diligent search himself in every likely place, and has kindly afforded me an opportunity for a personal examination this summer, we have not succeeded in finding their present quarters. It is possible these Bats were a colony of *V. daubentoni;* but it must be remembered that Mr Gray was well acquainted with that species, having reported its occurrence on several occasions, and that, after all, the existence of *V. nattereri* in the district is by no means so very unlikely, looking to its range on the Continent, and seeing it has apparently already occurred in Scotland, a specimen (an adult female) in the British Museum being labelled "Inveraray, August 1858," and identified by no less an authority than Dr G. E. Dobson ("Catalogue of the Chiroptera," p. 308). I ought to say, however, that the Duke of Argyll, from whom the specimen is said to have been received, has no recollection of the matter (letter to Mr Harvie-Brown, 24th March 1891). This record appears to have escaped the notice of Mr Alston when he drew up his paper on the Scottish Mammals, and might have been overlooked by me also, but for Mr Harting's article on the species, published in the "Zoologist" for 1889, p. 247.

To the same writer's article on the Whiskered Bat (*Vespertilio mystacinus* Leisl.) in the "Zoologist" for 1888 (p. 165), I am indebted for a clue which has enabled me to trace an undoubted Scotch example (the only one on record) of this species also. The specimen, which is in the Manchester Museum (Owens College), was captured by Mr J. Ray Hardy of that institution, who writes me as follows:—" I took the Bat you mention about four miles from Rannoch on the road to Pitlochry, early in June 1874, while sugaring for *Noctuæ*. I struck at him with my entomological net, and the cane rim caught him and knocked him down. He died in my hand."

Its identity with *V. mystacinus* of Leisler was first recognised by the Rev. J. E. Kelsall, who mentioned it to Mr Harting, and this identification has been confirmed by Mr Oldfield Thomas of the British Museum, to whom the specimen has been kindly submitted by my friend Mr W. E. Hoyle, curator of the Owens College Museum. My best thanks are due to Mr Hardy for the privilege of being allowed to record these particulars for the first time. This, then, is another Bat which we may reasonably hope to discover in the district.]

Order INSECTIVORA.

HEDGEHOG.

ERINACEUS EUROPÆUS *L.*

In spite of the persistent persecution to which it is subjected by gamekeepers in consequence of the occasional plunder of a pheasant's or a partridge's nest, this interesting animal is still common in all but the most unsuitable localities. Many of them come annually under my own notice between April and October—especially in those years when I happen to be much about the woods and hedgerows at night after moths. Some idea of their numbers may be gathered from the fact that a keeper on a small property a few miles south of Edinburgh kills between twenty and thirty annually. I have frequently kept Hedgehogs in confinement, but cannot say that they have always proved "interesting pets." The facility and speed with which they follow up the track of a beetle shows that they possess a very keen scent. Pale or albino examples occur at rare intervals—two (adult and young) belonging to the Earl of Haddington were exhibited at a meeting of the Royal Physical Society on 17th February 1885.

Pennant, in Lightfoot's "Flora Scotica," published in 1792 (vol. i., p. 13), says of the Hedgehog—"not found beyond the Tay, perhaps not beyond the Forth;" but the accuracy of this statement may well be questioned. Sibbald includes the "*Erinaccus*" in his "Historia Animalium in Scotiâ" (1684), but the few remarks he makes concerning it have reference entirely to its habits and uses. We may fairly assume, however, that had it then been unknown, or even very rare in Fife—a county whose animals were probably better known to him than those of any other part of Scotland—he would have made some allusion to the fact. Only six years after the publication of Lightfoot's work, the "Urchin" was included without comment in an enumeration of animals found in the parish of Dowally, near Dunkeld ("Old Statistical Account of Scotland," vol. xx., p. 472). It is also mentioned in the Account of the parish of Tillicoultry, written in 1795 (*op. cit.*, vol. xv., p. 200). Don, in his "List of Forfarshire Animals," published in 1813 (Headrick's "Agriculture" of the County, Appendix, p. 38), says: "This animal was formerly rare in Angusshire, but of late years it has appeared in tolerable plenty." From personal inquiries made in different parts of the counties of Fife, and Perth as far north as the entrance to the Highlands, I learn that it is common throughout these districts, and none of my correspondents can remember when it was otherwise. Mr Keay, gamekeeper, Murthly, can speak from his own knowledge to its abundance in that neighbourhood for over forty years. Sixty-three years ago, the limit set by Fleming to its northern distribution in Britain was the Moray Firth ("British Animals," p. 8).

COMMON SHREW.

Sorex vulgaris *L.*

Very abundant and apparently universally distributed in the district, expresses no more than the bare truth with regard to this species. Though so common, comparatively few people would be aware of its presence but for the feeble cheep and rustle in the grass, and the occasional dead body on the pathway, so seldom does the tiny creature expose his velvet coat to view. These indications of the animal's presence, however, scarcely give an adequate idea of its abundance, and it is only after we have had recourse to trapping for a time that this is fully realised. Wherever my traps have been set, from the vicinity of the seashore to the midst of the hills—whether by a stream, a hedge-bottom, or under a whin-bush; in a plantation, a garden, or an upland pasture—the Common Shrew has invariably been one of the first and most frequent captures. On the furze-clad slopes of the Braid Hills they are a perfect pest, continually occupying the traps to the exclusion of better things. In the heart of the Pentlands, too, near Loganlee, they have more than once frustrated my endeavours to obtain other kinds; and Mr Bruce, gardener, Colinton House, to whom I am indebted for examples of most of our smaller mammals, takes large numbers in and about the garden there. Correspondents in East Lothian, Peeblesshire, Linlithgowshire, and Fife, have had no difficulty in procuring me specimens. No better time for trapping them can be selected than during winter, especially when there is frost and a sprinkling of snow on the ground; and I have captured them in the daytime as readily as at night. Though probably most active towards evening and after nightfall, the Shrews cannot properly be regarded as nocturnal animals, nor do they appear to hibernate even partially.

LESSER SHREW.

Sorex minutus *L.* = Sorex pygmæus *Pall.*

The authors of the last edition of Bell's "British Quadrupeds" (1872) were disposed to regard the Lesser Shrew as generally distributed in Scotland; but, while the correctness of their assumption need not be questioned, it should be remembered that it was based on very scanty data; and it is to be regretted that during the nineteen years which have since elapsed our knowledge of the animal's actual distribution in the country has received scarcely any substantial increase. By the uninitiated the Lesser Shrew is hardly likely to be distinguished from the common species; but our present ignorance of its precise range north of the Tweed is not creditable to Scottish field-naturalists.

During the winter of 1888-89, I gave the ferryman at Cramond a few traps, which he set in and about his garden, on the Linlithgow side of the Almond. Among other things, he captured three examples of this tiny quadruped, one of which—the first recorded from this district—was exhibited by me at a meeting of the Royal Physical Society on 20th March 1889. In the course of last winter (1890-91) three or four others, captured by the Messrs Campbell a little farther west in Dalmeny Park, have passed through my hands; and from what Mr M'Leish, mole-catcher, Millburn, near Corstorphine, tells me, there can be no doubt he has observed it in his neighbourhood. On 22nd November 1890 Mr Eagle Clarke captured one in the daytime on the northern slopes of the Pentlands, at Colzium, as recorded in the "Scottish Naturalist" for January 1891, page 36; and on 25th February Mr T. G. Laidlaw brought me another which his brother had trapped the previous day at Hallmyre, near West Linton, Peeblesshire. According to Alston ("Fauna West of Scotland," p. v), it is not uncommon in the Upper Ward of Lanarkshire.

As these pages are passing through the press, I learn from Mr Eagle Clarke that a specimen has been sent to him by Mr Wm. Berry of Tayfield, Newport, who captured it on 2nd November (1891) on Tentsmuir, Fife.

WATER SHREW.

CROSSOPUS FODIENS (*Pallas*).

With us the Water Shrew is widely, though somewhat locally, distributed, but is nowhere abundant. I have records of its occurrence during the last four or five years in several parts of East Lothian and Midlothian; also in Linlithgowshire, Stirlingshire, Peeblesshire, and Fifeshire, and it has been observed by Dr Hardy to enter his own house at Oldcambus, Berwickshire ("Proc. Berw. Nat. Club," viii., p. 527). Through the attention of the Messrs Campbell, I have recently had opportunities of examining several in the flesh, captured both in summer and winter in Dalmeny Park, and have found them all to be more or less of the typical form with the light underparts, which is the common form in the more inland localities also. Mr Bruce has taken it in the grounds of Colinton House, at a considerable distance from water. One found dead by my children on the path close to the Braid burn at Greenbank farm, on 10th July 1890, was of the variety with the dark underparts—the *Sorex remifer* of MacGillivray's "British Quadrupeds." I have obtained several by means of the "Cyclone" traps baited with cheese.

Alston (Scottish Mammalia, p. 10) gives the credit of adding the Water Shrew to the Scottish list to Dr Scoular of Glasgow. As long ago, however, as 1808 it was known to Patrick Neill as an inhabitant of the Esk at Habbie's Howe, near Carlops (*vide* his list of animals and plants, contributed to the 1808 edition of Allan Ramsay's "Gentle Shepherd," vol. i., p. 269); and in 1812 Fleming, in his "Contributions to the British Fauna," published in the Wernerian Society's

"Memoirs" (vol. ii., p. 238), stated that it was then "by no means rare in the county of Fife." Its presence near Abbotsford was recorded in the "Magazine of Natural History" (iii., p. 236) prior to the appearance of Scoular's note. What Scoular (*op. cit.*, 1834, vi., p. 512) really did was to recognise the variety *remifer* for the first time as Scottish.

Since this memoir was read, I have had an unusually good opportunity of studying the habits of this interesting animal in the Braid burn below Comiston farm. About 8 P.M., on 22nd May, while strolling quietly by the side of the stream, a series of ripples spreading over the water from the bank almost beneath my feet attracted my attention. In a few seconds the wavelets had vanished, and the surface of the pool was as still as before; but while I gazed into it a little creature, clothed as it were in silver, darted from the bank to the bottom of the stream, and after hastily snatching some water insects or crustaceans from a piece of wood lying on the mud, returned precipitately to its den. In this way it continued to feed for some minutes, when the sudden appearance of three others swimming out from the opposite bank was evidently the signal for play, for in an instant the four joined company and scampered up the stream after each other with astonishing rapidity, swimming on the surface or beneath it, or running on the margin with equal facility. Having gone a distance of twenty-five or thirty yards, they disappeared into a drain, but soon reappeared, and proceeded down stream, in the same manner as they had gone up, till about twenty yards below where I stood, when they disappeared a second time. In a few minutes they were out again, and so the chase went on for fully half an hour. They frequently made a squeaking noise, which seemed to me identical with that uttered by the Common Shrew. With one exception they were quite pale on the under parts. One of them now forms part of the collection of native mammals in the Museum of Science and Art; and the gambols of the survivors have been a source of pleasure to me on several subsequent occasions.

MOLE.

Talpa europæa L.

Innumerable colonies of Moles inhabit all our cultivated lands and pastures, from the shores of the Forth to the summits of the hills; I have myself seen their "hillocks" at fully 1700 feet on the Pentlands, and still higher on the Ochils. In the lowlands, where agriculture is at its height, the farmers wage an incessant war against it through the medium of the professional mole-catcher, but in the upland pastures it is less molested, and consequently its habits and economy can be there more readily studied. In such outlying districts, on the lower slopes of the Pentland and Moorfoot hills, I have frequently dug into the large mounds, or fortresses as they have been called, containing the snug beds of soft grass in which the animals repose during the short intervals from labour in their subterranean hunting-grounds, but cannot say that I have found the number or position of the tunnels with which these mounds are pierced, disposed with the mathematical exactitude invariably ascribed to them on the strength of Le Court's observations. Occasionally a close agreement with the well-known illustration is observed, but as a rule the departure from it is very considerable. The following figure accurately represents the plan of a "fortress," about three feet in diameter, which I dissected in April 1891 on a moor behind the Dalmahoy hills. I ought to say that my attention was first directed to this want of uniformity by Mr M'Leish, mole-catcher, Corstorphine.

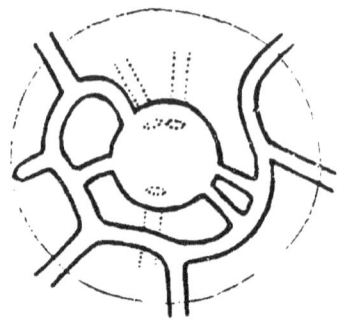

Under gallery communicating directly with the central chamber and the outside runs. Three escape-holes lower in the chamber are shown in dotted lines.

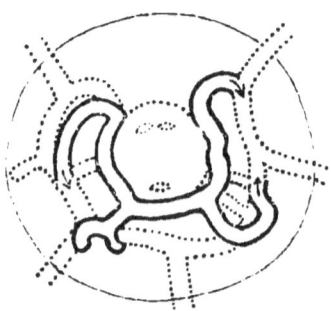

Upper gallery opening downwards (at the arrows) into the under one, which is shown in dotted lines.

The voracity of the Mole is well known. A year or two ago I placed one at dusk into a deep box in which a quantity of earth had stood for a considerable time. An hour afterwards, on looking into the box by the light of a lantern, innumerable worms were observed all round the sides endeavouring to make their escape. The Mole, however, was out of sight, but his presence was indicated by slight upheavals of the soil. By next morning he had not left a worm, and so keen was his appetite that it was found impossible to meet its demands, and he died after four days of confinement. Instances of Moles taking earth-worms from the hand immediately after being captured have been related to me, and incidents of a like import are recorded in "Some Observations on the Natural History and Habits of the Mole," by the Rev. James Grierson, M.D., minister of Cockpen ("Mem. Wern. Soc.," iv., p. 218), published in 1822.

In August last, while searching for land-shells in a birch plantation, a strange throbbing sound—a kind of thurr thurr,

thurr thurr—somewhat like the purring of a cat or the distant jarring of a goatsucker, arrested my attention. Creeping cautiously in the direction whence the sound came, I perceived that it proceeded from the ground, and the moment I touched the spot it ceased. On digging into the ground a Mole's run was found a few inches below the surface, and I have not the slightest doubt the author of the sound was no other than the Mole itself.

I have obtained the young (probably ten days to a fortnight old) from the nest during the third week of May. They were then of a bluish-grey colour, very silky in appearance, and without fur.

Buff and cream-coloured varieties are not very uncommon, and from time to time afford material for a correspondence in the newspapers.

Order CARNIVORA.

WILD CAT.

Felis catus *L.*

Though doubtless once a denizen of all our glens and deans, wherever the banks were sufficiently rocky and clothed with woods and thickets to afford the necessary shelter, this fine animal is now quite extinct throughout the whole of the district, and for many miles beyond it. Its extermination throughout Fife and the Lothians, with the exception of the eastern corner of Haddingtonshire, must have taken place a long time ago, but how long it is impossible to say, as no facts bearing on the point appear to have been placed on record. In some of the many suitable localities that readily suggest themselves—for instance, on the banks of the Esk, to go no farther afield—it is just possible that a few may have lingered till the opening years of the present century, but this is a mere conjecture on my part. In the cleuchs and deans of the Lammermoors, the adjoining coast of Berwickshire, and in the Border counties generally, it seems to have maintained its footing longer, and certainly existed in a few spots well into this century. The same may be said of the hills of Stirlingshire; and among the wilds of south-western Perthshire it did not finally disappear till some twenty-five or thirty years ago. In 1849 and 1874, Dr James Hardy, of Oldcambus, published, in the "Proceedings" of the Berwickshire Naturalists' Club (vol. ii., p. 357, and vii., p. 246), some valuable observations on the former occurrence of the Wild

Cat in Berwickshire (and adjoining parts of East Lothian) and the other Border counties—a timely step, otherwise even the memory of it would, in all probability, have died out without notice there as elsewhere in the south-east of Scotland. What Dr Hardy did for the Border counties, Mr Harvie-Brown has done for the rest of Scotland in his excellent article on the species contained in the "Zoologist" for 1881, p. 8. Dr Hardy's main facts will be best given in his own words. In 1849 he wrote as follows:—" The Wild Cat is probably by this time considered as an extinct animal in Berwickshire. According to my information, it has not been noticed in this part of the county for at least forty years. I have, however, recently ascertained that one at least yet survives, having hitherto been secured amidst the fastnesses of our rocky coast from the unremitting persecution waged in modern times against our indigenous wild Carnivora. On the 17th of March 1849, while on a visit to the coast immediately to the east of St Helen's Chapel, I had the pleasure of seeing an individual still frequenting the ancient haunts of its race. . . . I first remarked it on the top of one of these precipices named the Swallow Craig. . . . This was likewise the spot where, more than forty years ago, my father used to see them when they were still numerous. . . . The dark caverns, or 'coves,' of which there are several in the range of cliffs from this to Fast Castle, had the repute in former times of being tenanted by these animals. . . . By their occasional depredations in the hen-roost they were known as far westward as Dunglass, perhaps finding a retreat in the deep and wooded glen. Fifty years ago, they were exceedingly numerous in the woods above the Pease Bridge." The precipitous sea-banks between Gunsgreen and Fairneyside are mentioned as another haunt, and it is further stated that below a place named Blaikie, "there are several holes in the banks still called the cat-holes, which were the headquarters of the wild cats that prevailed there. . . . It is only within a recent period that the last of them was killed." In his 1874 article Dr Hardy adduces further evidence of the animal being formerly familiar to the country people in the Border districts. Alexander Somerville's encounter with it in the Ogle Burn, a deep dark wooded ravine running into the Lammermoors in the parish of Innerwick, as related in his "Autobiography of a Working Man," is quoted at length. Other localities particularised are "The Sting," on Upper Monynut, also in the Lammermoors, where the last "clecking" is said to have been destroyed about sixty years since; Belton Wood, where one is supposed to have been seen somewhat later; and the Press Woods, on the edge of Coldingham Moor. Place-names such as Cat-craig, now quarried away for lime, on the coast east from Dunbar; Wulcat Yett, a few miles from Jedburgh; Cat-lee Burn, in Southdean, are put forward as additional evidence; and in this connection I may also mention Wul-cat-brae on the Eye (*op. cit.*, ix., p. 15), and Cat-slack in Yarrow. In the "Old Statistical Account" (xvi., p. 76) the Wild Cat is included among the quadrupeds then inhabiting the parish of Castletown in Roxburghshire, a record which is not referred to either by Dr Hardy or Mr Harvie-Brown. In the tenth volume of the "Proceedings" of the Berwickshire Club (p. 47) it is recorded that at the Jedburgh meeting in September 1883, "a Wild Cat (*F. catus*) shot a few years ago at Wolflee," was exhibited by Mr J. T. S. Elliot; but I would ask,—Is there no likelihood of this having been a domestic cat run wild?

The "Old Statistical Account" contains abundant evidence of the Wild Cat in Stirlingshire in the end of last century. In the "New Statistical Account" of the county it is spoken of as extinct in 1842 in Campsie and Fintry, but as still existing in Strathblane, which, however, Mr Harvie-Brown is inclined to doubt ("Zoologist," 1881, p. 15). Cat-craig and Cats-cleuch, near Denny, are mentioned as probable place-names in Stirlingshire. Passing to south-west Perthshire, there is no lack of evidence of its presence in many localities there during the first half of the present century; but I must refer my readers to Mr Harvie-Brown's admirable paper for the details. Suffice it to say, that two were killed near Aberfoyle about 1855; that the last obtained in the Callander district was trapped in or about 1857 in the glen of Leny, and is still preserved at Leny House; that another was killed at Cromlix, Braes of Doune, in 1857 or 1858; that about 1850 one was killed at Gleneagles; that the keeper on Balquhidder killed Wild Cats about 1855; and that in the district south of Glendochart the last was killed upon Ben More, near Suie, in 1863 or 1864.

A few of the places mentioned above are rather beyond the limits of this paper, but their bearing on the subject is sufficiently obvious, it is hoped, to justify the reference to them.

FOX.

CANIS VULPES *L.*

Notwithstanding its predatory habits, the Fox is still fairly numerous, being allowed in most parts of the district a large amount of immunity from indiscriminate persecution in order that it may be hunted with hound and horn in orthodox fashion; otherwise, we may well suppose, it would have shared the fate of the other larger Carnivora, and long ere this have been practically banished from the lowlands. In the words of the writer of the article on the Fox in the Badminton

Library ("Hunting," p. 63), "hunted he must be; if he is to exist at all, it is his *raison d'être.*" At the present time there are ten packs of fox-hounds in Scotland, all located south of the Firths of Tay and Clyde. Six of them hunt the eastern division, namely—the Fife hounds, 50 couples (kennels at Harleswynd, Ceres); the Linlithgow and Stirlingshire, 35 couples (Golf Hall, Corstorphine); the Berwickshire, 35 couples (Belchester, Coldstream); the Duke of Buccleuch's, 57 couples (St Boswell's, Roxburghshire); Mr Scott-Plummer's, 20 couples (Sunderland Hall, Selkirk); and the Jedforest (Lintalee, Jedburgh). During the season 1890-91 the Linlithgow and Stirlingshire pack killed 24½ brace of foxes, as I am informed by Mr E. Cotesworth, the huntsman, who adds that the yearly average is about 25 brace. Mr W. Shore, the huntsman of the Duke of Buccleuch's pack, tells me that they usually kill about 30 brace in a season. From these statements I estimate that the six packs kill over 250 foxes per annum. Of course, a good many more are quietly got rid of in less demonstrative ways, even in the heart of the hunting areas; and in the hilly districts the keepers and shepherds openly capture or destroy all they can. Live cubs are readily disposed of at from 10s. to 15s. a-piece, to be turned out in hunting districts, chiefly in England. In the spring of 1889 a litter of five was dug out of an earth on the Pentlands above Dreghorn. A vixen and her six cubs, taken on the Peeblesshire hills in the end of April last, were sold at 10s. a-piece, while another vixen and five cubs, captured on the Pentlands above Boghall on 11th May, were disposed of at £3, 10s., being £1 for the mother and 10s. for each of her young ones.

Though, as has just been shown, this animal is by no means rare with us, it is comparatively seldom that a person not having special facilities has the opportunity of writing "Fox seen" in his diary. Still, in the course of the last fifteen years, I have observed them on many occasions (and smelt them on many more!) during my natural history rambles in Midlothian and the adjoining counties. Quite recently I had an excellent view of one on the Pentlands as it left the rocks above Swanston and trotted leisurely over the summit of Cairketton hill. When my father tenanted the farm of Tynefield in East Lothian, a litter was reared there every year; and I well remember the delight with which we used to watch the youngsters as they played at the mouth of the earth.

In the volumes of the "Old Statistical Account," the Fox is perhaps more frequently mentioned than any other wild animal. The writer of the article on the parish of Bowden (Roxburghshire) tells us (vol. xvi., p. 239) that "much [injury] was formerly sustained from foxes, to which the furze and brushwood on the lower skirts of Eildon, both in this and Melrose parish, afford cover. Of late, however, their number has been diminished by the noblemen and gentlemen of the Caledonian Hunt and others who keep hounds." In the Account of Duddingston (vol. xviii., p. 374), it is recorded that "Foxes from the neighbouring hill or plantations sometimes invade the farm-yards." Stark, in his "Picture of Edinburgh" (6th ed., 1834, p. 322), states that it is "occasionally seen on the southern declivities of Arthur Seat hills," a locality in which I have good reason to believe it has been observed up to a much more recent date. Mr Harold Raeburn tells me that his brothers have seen one well within the city boundary, near the Dean, within the last three or four years.

Such place-names as Todholes near Balerno, Todhills near Dalkeith, etc., perpetuate the old Scotch name for the Fox.

OTTER.

Lutra vulgaris *Erxl.*

The Otter occurs permanently or at intervals on all our rivers and larger streams, but only in very limited numbers. Without attempting to give an exhaustive list of localities and occurrences, I may mention that I have on several occasions seen their footprints or "seals" on the banks of the Biel burn in East Lothian, and only the other day by the Esk within the deer-park at Dalkeith Palace; and that, besides specimens killed in these places, I have, during the last few years, either examined examples or had their occurrence reported to me from the Tyne; the South Esk, on which one (of two) was captured near Dalhousie Castle in 1889; the North Esk, on which one was killed near Eskbank in 1890, and another seen at Newhall five or six years since; Glencorse reservoir and Logan burn, in the Pentlands, where one was captured in 1886, and the marks of another seen the winter before last; the Tweed, between Peebles and Innerleithen, and at various other points in its course; the Almond, both near its mouth and higher up; and the Carron, in Stirlingshire. I have also quite recently seen one from near Callander, and J. Gilmour, Esq. of Montrave, informs me it is still not uncommon in Fife.

From Sir Robert Sibbald we learn that in the end of the seventeenth century, when he wrote his "History of Fife and Kinross," "the sea-otter, which differeth from the land-otter, for it is bigger, and the pile of its furr is rougher," inhabited the Firths of Forth and Tay (1803 ed., p. 114); but I am not aware that the Otter now occurs in the waters of these arms of the sea. In the volumes of the

"Old Statistical Account" the Otter is frequently mentioned. For instance, in vol. xx., p. 49, we are told they abounded at Loch Mahaich, near Doune. In vol. xviii., p. 374, it is stated that they "used to frequent Duddingston Loch," which they have often visited since—see "Proceedings" of the Royal Physical Society, vol. i., p. 269, and vol. ii., p. 244, where mention is made of specimens obtained or seen near Peffer Mill in December 1856, Duddingston Loch about the same time, and near Coltbridge in December 1860. Mr Speedy assures me that not more than twenty years ago they regularly frequented the policies of Duddingston House. Patrick Neill, in his list of Habbie's Howe animals (1808), enters the Otter with the remark, "seldom met with" against it; and we find the fact of one being killed during severe weather in December 1812 at the farm-offices of Ingliston, a mile and a half from the Almond, considered worth recording in the "Scots Magazine" for that year (p. 892). A male killed near Stow in the end of 1831, and a female in November 1832, are recorded in the "New Statistical Account" of the parish (p. 404). Both were sent to the museum of the University of Edinburgh. The animal is also mentioned in Stark's "Picture of Edinburgh" (1834) as inhabiting the Water of Leith, "but is rare." On the whole, I am inclined to think that its status in the district now is not much worse than it was three-quarters of a century ago.

MacGillivray, in his "British Quadrupeds," 1838, p. 180, states that a pack of otter-hounds was then kept by Lord John Scott, who "exercised" them on the streams of Roxburghshire. Since the death, about nine years ago, of Mr W. Hill, who resided for some time at Kilduff, in East Lothian, where he kept a pack, I am not aware that otter-hunting has been practised in the district, except when the Dumfriesshire hounds pay the Tweed or the Esk a visit, which they have done quite recently. Mr Chouler, keeper, Dalkeith Park, has the head of the last otter killed by Mr Hill; it was taken in the Esk on 10th October 1881.

BADGER.

MELES TAXUS (*Schreb.*).

That the Badger, or Brock, as it was called, was a common animal throughout the district in olden·times goes without saying. At the time the "Old Statistical Account" was drawn up—the closing years of last century—it was still well known as an inhabitant of many localities, though even then its numbers were greatly reduced; and the adverse conditions continuing to grow, its extermination in most of its former haunts was apparently accomplished by about the middle of the present century. Here and there a miserable remnant lingered a few years longer, but it is very doubtful if more than eight or nine pairs of the original stock now exist anywhere in the valley of the Forth, and these mainly in its remotest parts among the Perthshire hills, concerning which the Rev. P. Graham wrote in his "Sketches of Perthshire" (2nd ed., 1812, p. 216), "We have hares, badgers, weasels, etc., everywhere." In the valley of the Tweed it maintained its footing better, and a few favourite habitats are known to be still occupied.

In the "Old Statistical Account" of Duddingston (vol. xviii., p. 374) we read that "a solitary badger at times may provoke a stubborn chace and contest," and it is interesting to know that at the present moment a few are to be found within a very short distance of that locality, though I fear we cannot claim them as the descendants of the sturdy beasts just mentioned. I refer to the policies at Edmonstone House, where Badgers have taken up their abode for some years past, and are known to have bred on several occasions. Unfortunately, the gamekeeper seems to think they are already too numerous, and has taken to killing them. In May of this year (1891) I saw two of them in the taxidermist's hands. It is supposed that this colony originated with a female which escaped from the stables at The Inch, where Mr T. Speedy has kept several in confinement. The Badger seen in a field near Greenend in June 1883, and mentioned in the "Scotsman" at the time, was doubtless the same animal.

Former haunts on both branches of the Esk have been placed on record. In 1808 Neill included it in his list of animals inhabiting the grounds of Newhall, on the North Esk; and the writer of the "New Statistical Account" of the parish of Borthwick, on the southern branch, informs us that while he was preparing that account (in 1839) there was a litter of young Badgers in the Chirmat, a piece of wooded hill opposite the windows of the manse. About Temple and Rosebery, in the same neighbourhood, it existed until quite recently, and the last may not even yet have been destroyed there. One which was taken alive near Temple was advertised for sale in the "Scotsman" of 25th April 1880; and a little farther east, on the confines of Midlothian and Haddingtonshire, another was trapped some sixteen or seventeen years ago at Blackshiels by Mr W. Wood, gamekeeper, who has often related the circumstance to me.

Almost every estate in East Lothian appears to have contained Badgers at one time. Mr Saunders, gamekeeper, Gosford, informs me that it is now some forty-five years since the last was killed there, and that about the same time they were on the adjoining properties of Gilmerton and Luffness, on the latter of which the last succumbed about thirty-seven years ago. From Mr R. Inglis, keeper, Tyninghame, I learn that about forty years ago he knew of two litters on that estate in each of two or three successive seasons, "but they were never allowed to live long." The last was killed there twenty-one years ago, and he is not aware of any having been seen since. There are places, however, in Binning Wood, where they may well have lingered some years longer. In May 1881 a hole once frequented by them in the grounds at Belton was pointed out to the members of the Berwickshire Naturalists' Club, and Dr Hardy was told that they were then preserved in Pressmennan woods, and retained a privileged home at Newbyth (Club's "Proc.," ix., 427). The latter part of this statement is not corroborated, however, by the Newbyth keeper, of whom I have made inquiries. He has been seventeen years on the estate, and has not known of a Badger on it during that time. Pressmennan woods, on the other hand, not only were, but I have good reason to believe still are tenanted. In 1862 I saw one alive in East Linton, which had just been brought in from the Biel estate, of which Pressmennan is a part; and Mr G. Muirhead tells me he has a specimen which was captured at Salton in the spring of 1868. On the confines of East Lothian and Berwickshire, as well as throughout the latter county, they had many haunts, and in few districts have more places been named after it. We have, for instance, the Brock or Spott water, near Dunbar; the Brock-holes, a bank below Thurston Mains; Brockhole farm, on the Eye; etc. (*vide* "Proceedings" of the Berwickshire Naturalists' Club, ix., pp. 17, 215, 222). Proceeding up the valley of the Tweed, we find it in Lauderdale even at the present time. It was stated to be still common at Legerwood in 1880 (*op. cit.*, ix., p. 242). During the last five or six years I have examined in the flesh about a dozen from within the watershed of the Tweed, most of them having been captured in the neighbourhood of Lauder, and two in Selkirkshire. On one occasion, some two-and-twenty years ago, Mr Small, taxidermist, Edinburgh, received six from Coldstream. In the parish of Heriot they were present in the days of the "Old Statistical Account" (xvi., p. 51); and in the north-west of Peeblesshire one was killed at Halmyre dean about ten years ago, as I am informed by Mr T. G. Laidlaw; while on the Dolphinton estate a full-grown female—the third got there during the last two or three years—was captured on 18th April 1890, as recorded by Mr Charles Cook in the "Scottish Naturalist" for January 1891, page 36.

The rough braes of many a Linlithgowshire stream, covered as they then would be with natural wood and bracken, were doubtless in former times also the chosen abode of the Brock; indeed this is rendered certain as regards one section of the county at any rate, by the fact that the parish of Uphall was formerly called "Strathbrok" ("Old Stat. Acc.," vi., 543); hence also Broxburn, the principal village in the parish. The Rev. Professor Duns informs me that when he went to reside at Torphichen in 1844, there were still a few in that neighbourhood, and that he has a skin yet which he then obtained. Lochcote was a habitat at that time, or even later, as I learn from the son of a former keeper there. Mr S. Martin, for many years keeper at Hopetoun, writes me (October 1891) that to the best of his recollection Badgers were killed there about twelve or fifteen years ago; and Mr Small, taxidermist, Edinburgh, tells me that prior to 1875 or 1876 he frequently had Badgers to stuff from Linlithgowshire. One, which I saw in Mr Small's shop, was killed on 29th September 1887, at the Witch-craig, by the Linlithgow and Stirlingshire fox-hounds. In the summer of 1881 the Earl of Rosebery had a pair sent from the south of England and liberated in Dalmeny Park, but both are supposed to have wandered and been killed. In 1889, as I am informed by Mr Bruce Campbell, three others, also from the south of England, were introduced into the grounds at Dalmeny, where they have bred, and seem now to be fairly established. I saw their earth recently (April 1891), and was told that four of the animals were seen near it a short time before.

The hilly districts of Stirlingshire, on the Forth and Clyde watershed, would seem at one time to have been quite a stronghold of beasts of prey, including "two species of Badger, . . . the one somewhat resembling a sow, the other a dog"! ("Old Statistical Account," Campsie, xv., p. 322); and the abundance of the animal in Doune (on the north side of the valley) and the neighbouring parishes has already been cited (p. 11). Brocks-brae, in the parish of St Ninian's, is a Stirlingshire place-name. As might be expected, a few still exist in the mountainous country at the head of the Forth valley. On 17th April 1889, I examined a fine male from the neighbourhood of Callander, and a little farther off, in the braes of Balquhidder, one was trapped last winter (1890-91).

Fife, like the other counties, had its Badgers at one time too, but they must have been rooted out many years since. In the days of my father's boyhood, some sixty years ago, they had not ceased to exist in the woods at Dysart. Mr Gilmour of Montrave writes me that a Badger was got in Wemyss woods some years back, but he thinks it was an escaped one; and from Mr Charles Cook I learn that another was caught on Benarty hill "some few years" prior to 1880.

PINE MARTEN.

MUSTELA MARTES *L.*

Once common, and with practically the same distribution as the Wild Cat, the Marten seems to have been extirpated as a resident species in the district even earlier than that animal; but, being apparently more given to wander, it has since turned up at wide intervals in localities from which it had long disappeared as a resident. Now, however, that it is being daily driven farther and farther into the Highlands, the chances of such stragglers reaching us are becoming more and more remote.

To Dr Hardy, of Oldcambus, we are indebted for bringing together what little is known of the occurrence of the Marten in the south-eastern part of the district, and I think I cannot do better than quote his remarks as printed in the "Proceedings" of the Berwickshire Naturalists' Club, vol. viii., page 527:—"In the Statistical Account of the united parishes of Cockburnspath and Oldcambus, p. 299, prepared in December 1834, the Rev. Andrew Baird reports that the Marten (*Martes Fagorum*) is said, a good number of years ago, to have inhabited the woods near the Pease Bridge. Till lately I had supposed that this hearsay had originated from some traditions of the Wild Cats that once made those woods their rendezvous; but now I think its correctness is undoubted, as Mr Peter Cowe, of Lochton, has an actual specimen of the Marten to show, and had heard of another in the very locality that I had questioned. The one preserved in Mr Cowe's collection, he writes of date 27th March 1879, 'was caught in Dowlaw dean in 1862 in a rabbit-trap. I had it alive for a week, but it would not eat. A short time, say a few weeks, after, another was caught about the Pease Bridge, but was destroyed before I heard of its capture.' Mr Kelly records that a Marten was trapped in 1848 in Lauderdale, by Mr Scott, which was the only example known there for half-a-century. It was stuffed by Walter Simson, Lauder. This furnishes us with four Berwickshire instances of the animal."

For the rest of the district I have practically nothing to add to what is recorded in Mr Harvie-Brown's article on the species contributed to the "Zoologist" for 1881, p. 81. In the "Old Statistical Account" it is mentioned as occurring among the rocks on the Campsie Fells (vol. xv., p. 323), and in the neighbourhood of Callander (vol. xi., p. 598); and in 1838 MacGillivray described a specimen from Lanarkshire ("British Quadrupeds," p. 168). Since then examples have turned up unexpectedly in a number of places. Thus on 10th May 1870 a very fine male, which I had the pleasure of exhibiting at the meeting of the Royal Physical Society in April 1891, on behalf of its owner, Mr Charles Cook, was caught in an ordinary rabbit-trap on the wooded slopes of the East Lomond hill, near Falkland, Fife; and in the summer of 1873 another male was trapped in the woods at Broomhall, near Dunfermline, in the same county, by Mr Stark, gamekeeper there, in whose hands I have recently seen it. In the beginning of June 1879, Dr A. C. Stark had an excellent view of one in the fir wood behind Callander, as he informed me shortly afterwards; and in February of the same year one was killed in Glenartney. In April 1880 one which was killed in Balquhidder on the 2nd of the month, was exhibited at a meeting of the Royal Physical Society; and, so recently as August 1888, another (a male) was caught on Urie estate, near Stonehaven ("Scotsman," 1st September 1888).

POLECAT OR FOUMART.

MUSTELA PUTORIUS *L.*

Seldom have the processes of extermination worked more rapidly and effectually than in the present case. Formerly abundant and generally distributed in the district, the Polecat has for many years been practically extinct, even in the more outlying localities. From the day that steel-traps came into vogue for the capture of the Rabbit, the fate of the Polecat in the lowlands was sealed.

The frequency with which it is mentioned, without any qualifying remarks, in the volumes of the "Old Statistical Account" is excellent evidence that it was a common animal in many localities, if not indeed in all, up to the closing years of last century. For thirty or forty years more it was still well known, but its numbers had been terribly thinned in the interval, and by about 1850 it had practically ceased to exist within our limits, so that the subsequent appearance of an example here and there has always been regarded as an exceptional event, and very likely some of these have merely been escaped Ferrets. Even the memory of it is fast dying out, and comparatively few of the keepers I have questioned can give me any information regarding it. Neill, in his Newhall list (1808), and again in his Tweeddale list (1815), includes the "Polecat or fitchet" without remark, but neither Stark (1834) nor Rhind (1836) mentions it among the animals to be found in the immediate neighbourhood of Edinburgh; while MacGillivray, writing in 1838, speaks of it as "of rare occurrence in the more cultivated tracts." Gamekeepers who have known some of the largest estates in Midlothian and East Lothian for more than half a century, are nearly unanimous in fixing the date of their last Foumart somewhere between 1840 and 1850. According to Dr Crombie, one was shot about a mile from North Berwick about 1860, and the only Midlothian Polecats Mr Charles Cook has a note of are one obtained on the farm of Fala Hill, and one seen at Crosswood Hill, both a number of years prior to 1880. At Edmonstone, near Liberton, one, which was afterwards trapped by the keepers, was seen by Mr James Haldane about thirty-five years ago (*vide* Mr Harvie-Brown's article on the Polecat in the "Zoologist" for 1881, p. 161). In Linlithgowshire nine were killed at Lochcote between 1838 and 1845 by the keeper, David Kerr, whose son I have recently interrogated on the subject. In 1847 Kerr also killed two at Champfleurie in the same county; and Professor Duns tells me that shortly after he went to Torphichen, in 1844, he noticed Polecats nailed to a keeper's wall. In Fyfe's "Summer Life on Land and Water at South Queensferry" (1851), the following passage occurs at page 148 :—" Amongst the *feræ naturæ* of Barnbougle, or rather of Dalmeny Park, no rambler, gifted with the sense of smell, could possibly omit the fitchet, foumart, or polecat (*Mustela putorius*), one of our finest furred animals, which we have reason to judge must be abundant in these woods, although indeed a species of fungus is found in them, which might, from its alarming smell, be apt to mislead to the belief that a polecat was near." The qualification with which the author closes his remarks will, it is to be feared, render the rest of his statement practically worthless in the eyes of most naturalists.

Mr Sam Martin, for many years keeper at Hopetoun, writes me that the last was killed there fully thirty years ago. Mr Durham of Boghead, near Bathgate (son of the late Mr Durham Weir, MacGillivray's able correspondent), has recently shown me two stuffed specimens which were captured there about forty-five years ago, and he further assured me that one was seen at a farm close by so recently as 1884. But this is not the latest record, for Mr W. H. Henderson, Linlithgow, writing to Mr Eagle Clarke in November 1890, states that at Kinneil, on the western confines of the county, on the 15th of November 1886 one ran out of a covert into a whinny mound. Mr Henderson had not seen or heard of one before in the county during a residence of thirty-five years. In the east of Stirlingshire Mr Harvie-Brown is of opinion it cannot have been common even sixty to sixty-five years ago. About 1860 several are said to have been seen on Gallowmuir by an old mole-catcher, and Mr James Stirling of Garden heard of one near there in the winter of 1879-80.

For Perthshire there are many records during the last forty years, but very few of them fall within the scope of this memoir. At Leny, near Callander, one was trapped in 1855, and another in 1858, while on Lord Moray's estate above Doune one was caught about 1850. In Kinross one was seen at Turfhills about 1845, and another on Scotlandwell Moss about 1860. With regard to Fife, I have often heard my father relate his experiences when a lad in connection with the trapping of Polecats in a poultry-yard near Dysart; this would be between 1830 and 1835. According to Mr Harvie-Brown's information, none have been seen at Lathirsk since about 1860, and at Lawhill one was obtained in 1866. In 1880 one was often seen in the grounds at Falkland, and when chased it took refuge in the thick ivy of the palace walls; the keeper tried to trap it, but without success ("Zoologist," 1881, p. 166). Mr Gilmour of Montrave writes me that he has not heard of a Polecat in Fife since he went to reside there in 1873.

Within Dr Hardy's recollection they were plentiful in the east of Berwickshire, in such localities as Dowlaw dean and the Pease dean woods, but he has not seen one "nailed up" for a long time, though he believed it had not been extirpated in the former of these localities so recently as 1880 (Harvie-Brown in "Zoologist," 1881, p. 162). From Mr R. Inglis, keeper, Tyninghame, I learn that a brother of his killed a number at Dunglass about fifty years ago. Mr Thomas Hope, taxidermist, Edinburgh, tells me he has seen a good many killed in the neighbourhood of Jedburgh—the last about thirty years back; and in the Berwickshire Naturalists' Club "Proceedings" for 1883 (vol. x., p. 269), it is stated that the last in "Black Andros" wood in Yarrow was killed "some years ago."

WEASEL.

MUSTELA VULGARIS *Erxl.*

This, the smallest of our Carnivora, is also the commonest, being still fairly numerous and generally distributed. About the farms and plantations of the Lothians it is a familiar object, preying for the most part on mice and small voles, which I have frequently watched it capturing. The actions of the Weasel when driven from its prey are most interesting, but require to be seen to be properly appreciated. On 9th January 1886, I observed one crossing the path in Dalmeny Park with something dangling from its mouth. On my throwing a stone at it, it dropped the object—a pretty little Bank Vole—and darted out of sight among the rough herbage. Taking my stand within two yards of the dead vole, I had not many seconds to wait till the Weasel reappeared, now sitting bolt upright (its heart throbbing with excitement), now plunging out of sight again, or bounding along the bank and across the road to see if the lost dainty could not be more easily recovered from the opposite bank. Having repeated these manœuvres for some time, it at last made a bold dash at the vole, and would have carried it off, but for my interference. Occasionally, however, one sees manifestations of its cruelty that make us think of revenge. In June 1890, while sauntering along a secluded path in Gosford woods, I noticed a thrush's nest in a bush about five feet from the ground, and being curious to see what it contained, I proceeded to pull the branch on which the nest rested towards me, when out sprang a Weasel. In the nest were the mangled remains of several young mavises not more than five or six days old, on which it had feasted. Nevertheless, I would be extremely sorry to see so interesting a member of our *feræ naturæ* wiped out of our fauna. To the farmer it is an undoubted friend, and he should certainly be the last to lift a hand against it.

In 1888 Mr T. Speedy obtained from this and other parts of Scotland several hundred Weasels and Stoats for transportation alive to New Zealand, where they have been turned down in the hope that they may provide a natural remedy for the Rabbit plague in that country.

STOAT or ERMINE.

MUSTELA ERMINEA *L.*

In spite of persistent persecution, the Stoat is still by no means rare, though not so numerous as the Weasel. It is of course more confined to the hilly districts than that species, but I have met with it in the low country as well as on the hills, and in the plantations as well as in the open. When ferreting Rabbits on the wooded banks of the Esk above Penicuik, I have several times seen a Stoat bolt before the Ferret; and in the spring of 1888, while resting by the side of a fir plantation at Loganlee in the Pentlands, where numbers are trapped every season, I watched one climbing the trees and jumping from bough to bough almost as nimbly as a Squirrel. When pursued it leapt to the ground from a height of nearly ten feet, preferring evidently to make its escape on *terra firma*. The speed at which a Stoat can move along on the ground is astonishing.

During the winter months numbers are received by the Edinburgh taxidermists for preservation. They are then in the white or ermine state. Of between twenty and thirty examined by me during the winter 1890-91, only two or three—obtained near Lauder and Gorebridge in the end of January and February—had changed colour completely; all the others were more or less brown on the upper part of the head and neck, and many of them had also a dorsal line of the same colour, but much paler, owing to a large admixture of white hairs. By 26th March specimens in perfect summer dress began to come in.

GREY SEAL.

HALICHŒRUS GRYPUS (*Fabr.*).

This large Seal is well known in fluctuating numbers at the mouth of the Tay, whence Professor Turner has received specimens; and though I cannot point to a record of its actual capture in the waters of the Forth, there can be no doubt it has frequently visited, if indeed it does not habitually frequent, the seaward portion of that firth as well. The discovery of bones, identified by Dr M'Bain as belonging to this species, in a kitchen-midden on Inchkeith, proves it to have been an inhabitant of the firth in former times ("Proc. Soc. Antiq. Scot.," ix., p. 453).

So long ago as 1841, Selby recorded, in the "Annals and Magazine of Natural History," the plentiful occurrence of the Grey Seal on the Farne Islands off the coast of Northumberland; and the "Great Seal" of Don's List of Forfarshire Animals (Headrick's "Agriculture" of Angus, 1813, App., p. 37) is no doubt also referable to this species. The late Mr Robert Walker, in an interesting article on the species contributed to the "Scottish Naturalist" for 1875 (vol. iii., p. 158), expressed the opinion that it was then the Seal most commonly met with on the east coast of Scotland, but I scarcely think this is the case now, whatever it may have been at that time. "It may be seen," he says, "all the year through at the mouth of the Tay, and along by the Carr Rock chiefly in summer. In autumn they congregate in great force in the vicinity of the banks of the Tay. These banks form a favourite resting place for them when the tide is out, as many as twenty having been counted at a time. In 1863 six specimens of this seal were caught in the salmon nets at Tentsmuir, some of them large animals, and all more or less ferocious and difficult to secure. The largest example was estimated by the fishermen to weigh fifty stones."

Early in March 1870 an adult female, measuring 7½ feet in length and weighing 33 stones, was captured in Mr Speedie's stake-nets, at the Tentsmuir station, mouth of the Eden, near St Andrews, and secured for the Anatomical Museum of the Edinburgh University by Professor Turner, who gave an account of it in the "Journal of Anatomy and Physiology" (vol. iv., p. 270). According to Mr Walker ("Scot. Nat.," iii., 159), another was captured along with the above; and Professor Turner states that in the previous spring two young examples, captured in the salmon-nets near Montrose, were sent to the Anatomical Museum.

At the mouth of the Tay I have myself frequently seen large Seals, undoubtedly belonging to this species. In the autumn of 1886 I had an excellent view of one gamboling with its cub on a sandbank at the mouth of the Eden. Walker, it will be observed, notes it to the Carr Rock, which therefore gives it a place in the fauna of the Forth. I well remember the large number of Seals which, twenty-five to thirty years ago, annually appeared about harvest-time in the Tyne estuary near Dunbar, many of which, I am persuaded, belonged to this species.

[GREENLAND OR HARP SEAL.

PHOCA GRŒNLANDICA *Fabr.*

A young Seal obtained many years ago at the mouth of the Firth of Forth was somewhat doubtfully referred by MacGillivray to this species ("British Quadrupeds," 1838, p. 209). It does not appear to have been preserved, so that we have now no means of judging of the correctness of his surmise. In former years, when this Seal was more abundant in its northern habitats, it is not improbable that one or two may have occasionally wandered to our shores, but as our information at present stands, we would scarcely be justified in giving the species a full place on our list.]

COMMON SEAL.

Phoca vitulina *L.*

Although not so abundant as formerly, this is still a common animal in the Firths of Forth and Tay, where it may be seen all the year round. Off the southern shores of the Forth, from Dunbar to Prestonpans, I have watched them on many occasions, and I recently saw one living in confinement which had been taken in the salmon-nets at Dalmeny. But their headquarters appear to be on the north side of the firth westward from Aberdour, and in the bay above North Queensferry within the estuary proper. When boating among the islands off Aberdour during the summer months, I have invariably found them present, occasionally in considerable numbers. On New Year's day 1886, I discovered a small one, apparently asleep, on a rock in Dalgetty Bay. It being low water at the time, I was able to walk within 12 to 15 yards of the animal, whose slumbers I rudely broke by a thump on the ribs with a good-sized stone. Instantly bending itself like a bow, with the central part alone resting on the rock, it gave a sudden jerk and sprang into the water. Once there it evidently considered itself safe, and, reappearing about 20 yards farther off, gazed in astonishment at the cause of the sudden interruption.

The Du Craig, a small islet off Rosyth Castle above North Queensferry, has long been noted as a favourite haunt of the Common Seal (*vide*, for instance, Fyfe's "Summer Life on Land and Water at South Queensferry," 1851, p. 270). When visiting this rock on 5th July 1884, for the purpose of identifying the terns which annually resort to it to breed, I noticed a number of Seals, some of which followed our boat at close quarters for a considerable distance.

The following extract from the Accounts of the Lord High Treasurer of Scotland in the days of James IV., shows that Seals then, as now, frequented the Isle of May, to which that monarch was a frequent visitor:—"1508 [8 Mar.] Item, that day to the heremyt of Maij that brocht ane Selch to the King . . . . xiiijs." (Stuart's "Records of the Priory of the Isle of May," p. lxxix). Sibbald, in his "History of Fife and Kinross" (1710), mentions the Seal several times. Many of the "Phoca, or Vitulus marinus, the Seal: our fishers call it a Selch," he says, "frequent the coasts of these two firths" (*op. cit.*, ed. 1803, p. 114); and, speaking of the Isle of May, he remarks that "many Seals are slain upon the east side of it" (*ib.*, p. 101). Quoting from a charter of David I. to the Monastery of Dunfermline, Sibbald further shows that Seals were a matter of trade in the twelfth century (*ib.*, p. 295).[1] In Stark's "Picture of Edinburgh" (1834), p. 322, we are told that "in the Firth of Forth the Seal (*Phoca vitulina*) is continually showing its black head;" and in the "New Statistical Account" of Alloa (1840), it is recorded that Seals "are constant inhabitants of the Forth here."

[1] The words of the charter are:—"Et de seliches qui ad Kingornum capientur, postquam decimati fuerint; concedo ut omnes septimos seliches habeant."

HOODED OR BLADDER-NOSE SEAL.

CYSTOPHORA CRISTATA (*Erxl.*).

One of the very few authentic instances of the capture of this inhabitant of high latitudes on the British coasts, is that of a young male taken opposite St Andrews on 22nd July 1872, and of which the late Mr Robert Walker, of the University of that town, gave a detailed description at the time in the "Scottish Naturalist" (vol. ii., p. 1). It was about 4 feet in length—47 inches was the exact measurement—and "when discovered it was reposing, near low-water mark, on the top of one of the ledges of rock that stretch out into the sea."

It has been suggested that the "sundry fishes of monstrous shape," mentioned by Boece, "with cowls over their heads like unto monks, and in the rest resembling the body of man," whose appearance in the Firth of Forth in 1577 caused such consternation among the superstitious folks of those days, may have been Hooded Seals.

Order RODENTIA.

SQUIRREL.

Sciurus vulgaris *L.*

At the present time the Squirrel is a common animal throughout the length and breadth of the district; indeed it is safe to say there is scarcely a wood of any extent in any part of it which does not contain at least a few. Yet it was not always so, for apparently the Squirrel was either entirely absent or very scarce in the south of Scotland when introduced at Dalkeith in the latter part of last century. Mr Harvie-Brown, who has made the history of the animal a special study, and, as the outcome of his investigations, has published a long and interesting paper in the "Proceedings" of the Royal Physical Society (vols. v. and vi.), considers there is no evidence of its prior existence in this section of the country, and lays little stress on the statement in the "New Statistical Account" of Berwickshire (1841, page 299), that "the Red Squirrel is said to have been at one time a denizen of Dunglass woods, in Cockburnspath parish." I cannot help thinking, however, that it must at one time have been indigenous in the Lowlands, and have gradually retired to the Highlands in consequence of the destruction of the ancient woods and forests; otherwise, what are we to make of Sibbald's statement ("Scotia Illustrata," 1684),—"In Meridionalis Plagæ Scotiæ Sylvis reperitur" (It is found in the woods of the southern part of Scotland)? There can be no doubt, however, that the south-eastern counties owe their present stock very largely, if not entirely, to the introduction of a few, first at Dalkeith Park about 1772, and then at Minto in 1827. The history of these introductions, and the subsequent spread of the species, are so very fully set forth in Mr Harvie-Brown's paper, that I need only refer to a few of the leading facts, and draw attention to one or two records which he does not allude to.

The current belief, from the time of the "Old Statistical Account" till now, is that Elizabeth, Duchess of Buccleuch (the present Duke's great grandmother), introduced Squirrels from England about 1772 to the menagerie which her husband (Duke Henry) then kept in the park at Dalkeith. Gaining their liberty either accidentally or by design, and finding a congenial home in the woods of the park, they increased with astonishing rapidity, so that in the course of the next twenty or thirty years they had spread eastward into Haddingtonshire and westward over the entire valley of the Esk. Here is what the minister of Pencaitland, in East Lothian, had to say of it in 1796: "The young woods on the estate of Fountainhall, it has been observed, have of late suffered much from Squirrels, which were introduced some years ago at Dalkeith, and have spread to this neighbourhood. They have attacked the Scotch firs in the proportion of about one in twenty, and almost every larix and elm. Already many of each of them are killed. If the harm they do in other places be as great, and be progressive as they multiply, this intended improvement will be unfortunate" ("Old Statistical Account," vol. xvii., p. 36). In 1791 it had "lately arrived at Penicuik from the menagerie of the Duke of Buccleuch" (*op. cit.*, vol. i., p. 132); and in 1795 the writers of the account of the parish of Glencross, of whom Professor J. Walker was one, record that "the Red Squirrel has become extremely common of late years. In this neighbourhood, the woods abound with them, and they are pretty numerous at Woodhouselee" (*op. cit.*, vol. xv., p. 439). Then in 1808 Patrick Neill records it for Newhall, which is much farther up the Esk, and where it had already given its name to the Squirrel's Haugh, adding, "introduced from England, but now common" ("Gentle Shepherd," i., pp. 270 and 279); and it is evidently the same naturalist who, in Pennecuik's "Tweeddale" (ed. 1815, p. 103), states that the animal was "Introduced on the North Esk, from England." This looks not unlike a separate introduction, but it may, of course, merely refer to the Dalkeith one. In the course of the next few years it had spread through Linlithgowshire into Stirlingshire, and even beyond the Forth into Clackmannan and South Perthshire,—where no doubt colonists from the north were met,—so that when the "New Statistical Account" was drawn up it was frequently alluded to. The colonisation of Fife seems to have been entered on somewhat later, and to have proceeded more slowly. Peeblesshire is also supposed to have been colonised from Dalkeith (the doubt expressed in Chambers's "History of Peeblesshire," Appendix, p. 525, is scarcely worth considering); but Roxburghshire, Selkirkshire, and Berwickshire are thought to have been stocked mainly from Minto, where several which the gardener there had obtained from Dalkeith in 1827 shortly afterwards made their escape. According to Dr Hardy, it appeared in Penmanshiel wood, in the east of Berwickshire, as early as 1830 or 1831; and 1838 or 1839 is the date fixed by Mr Kelly for their first appearance in Lauderdale, where they rapidly increased, and necessitated an order for their destruction in 1849, in consequence of the damage they were committing among the young fir trees ("Proc. Berw. Nat. Club," viii., p. 527).

The Squirrel has sometimes been accused of killing birds, merely because their bones have been found in its dreys, but as well might I argue that it occasionally kills sheep, because I recently observed one gnawing a shank-bone of that quadruped. Observing the little animal busy with something on the ground in a large fir wood, I walked towards it, when it at once scampered up a tall clean-stemmed tree, holding the object in its mouth. Having reached a branch about fifty feet from the ground, it sat down, and, grasping the prize between its fore-paws, began nibbling at the end of it. On my striking the branch with a stone, it dropped the object, which, to my surprise, was a sheep's shank-bone, measuring fully seven inches in length. A large hole had already been gnawed in the thick end of it.

According to Bell, the young are born in the month of June, and MacGillivray's statement is to the same effect; but I am inclined to think April is the more usual time, and that a second litter may frequently be born in the latter part of summer. Unfortunately I can only give one exact date, namely, 23rd April, on which day a nest containing several young Squirrels was discovered. At least two other instances, however, of young in April have come to my knowledge; also one in August. As to the supposed hibernation of the Squirrel, I can only say that I have seen them frisking about in every month of the year.

[DORMOUSE.

MUSCARDINUS AVELLANARIUS (*L.*).

In 1838 MacGillivray wrote, "This species . . . has not hitherto been satisfactorily proved to exist in Scotland, although it has been reported to me to occur near Gifford in East Lothian" ("British Quadrupeds," p. 236). No evidence in support of this statement has ever been forthcoming, and we must therefore conclude that his informant was in error.]

WATER VOLE.

ARVICOLA AMPHIBIUS (*L.*).

This well known and, for the most part, harmless creature, is abundant on the banks of all our streams, ditches, and ponds, where it may be constantly seen and its habits studied without difficulty. Any kind of country appears to suit it, so long as there is water at hand. It is equally at home, for instance, by the marshes on the coast, the ditches bordering the corn-fields, the ponds in the midst of plantations, or the burns meandering among the hills. It reaches a considerable elevation, for in May 1887 I saw a buzzard capture one on the hills above Loch Skene. When the bird had devoured it, I went to the spot and picked up the skin, which was so little damaged that it might very well have done for making into a stuffed specimen.

Occasionally this animal takes up its abode in our gardens, where it makes "runs" and commits considerable damage, destroying even shrubs and young trees by gnawing their roots. Several instances of this have come to my own knowledge. In March 1887 I obtained an old male from Dr Ronaldson's garden, Bruntsfield Place, Edinburgh, which had almost killed several bushes and young apple-trees by cutting off their roots. I handed the Vole and a specimen of its work to Professor Duns, who recorded the facts in a note which was published in the ninth volume of the "Proceedings" of the Royal Physical Society, p. 325. In a previous note bearing on the habits of this species (*op. cit.*, vol. v., p. 352), Professor Duns recorded the capture of another in an Edinburgh garden, where it had been feeding on beetroot.

The Water Vole is sometimes accused of killing young birds, and I am not prepared to affirm that it never does;

but I believe such an occurrence must be very exceptional. On Luffness marshes, where the animal is very abundant, I have seen a young redshank lying half-eaten at the mouth of one of their burrows,—no proof, however, that the Voles had killed it. My explanation is that, finding the bird dead, they were tempted to eat it, in the same way that Field Voles will devour a dead companion.

In the spring of 1890 a colony established themselves in a piece of rough, sandy ground by the public road near where a small stream enters the sea at Gosford Bay. For fully a month I passed the spot twice a day, and was much struck with the want of fear which they displayed, several always sitting unconcernedly about the entrances to their burrows while vehicles and pedestrians moved past within a few yards; indeed, so little notice did they take of people passing by, that Mr Eagle Clarke knocked one over with his walking-stick.

The black variety—the *Arvicola ater* of MacGillivray—is not common, but occurs from time to time in every county. I have notes concerning examples taken in Berwickshire, Roxburghshire, the three Lothians, Stirlingshire, Perthshire, and Fife. The Fife specimens, which as usual were small animals, were captured near Colinsburgh, where the form appears to be not uncommon. In the Highlands it is decidedly more numerous than in the Lowlands.

FIELD VOLE.

ARVICOLA AGRESTIS *De Selys.*

The Field Vole is abundant and generally distributed from the coast-line to the most inland localities, living among rough grass in meadows, young plantations, moors, and hill-pastures alike. Formerly I was in the habit of looking upon it as everywhere more abundant than the next species, but this view has not been borne out by my recent investigations. In the immediate neighbourhood of Edinburgh, for instance, I have trapped three *Glareolus* for one of *Agrestis*, and I am inclined to think that the former is likewise at the present time the commoner animal in many other parts of the fertile belt of country bordering the shores of the Forth, and probably the same may be said of the valleys of the Tay and the Tweed. But the moment we reach the hills and the moorlands, *Agrestis* becomes the commoner, and is in many districts apparently alone present. A number of years ago, when my home was at Macbiehill in Peeblesshire, it was very common there, and Mr J. Thomson, who has sent me a specimen, tells me it is abundant about Stobo in the same county. Within the last two years I have obtained very typical specimens from Aberlady, Dalmeny, Colinton, Dreghorn, the Braid hills, and the Pentlands. On the southern slopes of the Pentlands, near the farm of Boghall, there is a young fir plantation filled with tussocks of the *Aira cæspitosa* grass, and here *Agrestis* is in its element, burrowing under the tussocks, whose tender shoots supply it with abundance of food during winter and spring. By setting a few traps in the little "seats" at the mouths of the burrows, I have had no difficulty in capturing the inmates. Beds of *Juncus* also form favourite haunts. Though they certainly remain more at home in winter than in summer, they do not in any sense hibernate, and while they probably move about more or less at all hours, I am inclined to think they are most active towards evening. In winter afternoons I often see them about the entrances to their burrows. Owls and kestrels (to say nothing of weasels) of course destroy great numbers. Besides finding their remains in the "castings" of these birds, I have seen in the nest of a long-eared owl near Balerno several of this and other small rodents lying ready for consumption.

In the Southern Uplands the Field Vole, or Hill Mouse as it is there often called, at times multiplies to such an extent, and with such astonishing rapidity, as to assume the character of a veritable plague. The year 1876, for instance, was a memorable example. For a year or two previously they had been observed steadily increasing, no doubt in large measure owing to a succession of favourable winters, and reached a climax in 1876, when the pasture on whole hill-sides was destroyed by them. The country about Hawick seems to have suffered most. In the Borthwick-water district alone 10,000 acres of pasture were wasted to a greater or less degree—the damage being estimated at not less than £5000. A full account of this plague was prepared by Sir Walter Elliot for the "Proceedings" of the Berwickshire Naturalists' Club (vol. viii., p. 447). Sir Walter speaks of the present species only, but I imagine that in some localities at any rate the Bank Vole, which Dr Hardy (who identified specimens) tells us in a subsequent volume of the "Proceedings" (x., p. 278) was in great numbers at Faldonside in 1883, would also be present.

Since the above was written in April 1891, another Vole plague in the Border counties has become notorious. The subject is thus referred to in an article in the "Scotsman" of 12th November :—" Some three months ago reference was made to what is spoken of and felt as the mice plague on the Borders, and which was then affecting to a serious extent most of the farms in the western portion of Selkirkshire and the adjacent districts of Dumfries and Roxburgh shires. Since then there has been no mitigation of the pest, but on the contrary a great extension of the area over which it is spread, and an intensifying of its ravages. From inquiries made within the past week, it has been ascertained that the outlook, as the winter approaches, becomes more and more serious. The vermin have multiplied greatly during the summer, and they now swarm in numbers which defy computation. The high-lying farms on the western border of Selkirkshire seem to be suffering most. . . . Throughout the summer, grass and other herbage chiefly were preyed upon. The grassy farms have suffered, and are suffering most. The vermin do not seem to live on the lea grasses or dry hill-sides; the grassy bogs and white bent are the places where they abound most. Wherever the ground is what the shepherds speak of as 'not bare,' there they swarm in greatest numbers. They nibble and gnaw the long grass close to the ground, and the land is rendered altogether valueless for winter and spring feeding. Hundreds of acres of the best pasture land on many farms have thus for the present been totally destroyed, and whole hill-sides wear a blasted and desolate aspect, the ground being perfectly riddled by their holes and runs. In the autumn months hayricks were infested by the mice in countless numbers, and the hay has in many cases, as one observer expressively says, been minced into perfect chaff. Then the corn-stooks swarmed with them, as many as four or five nests being frequently found in a single sheaf. Now they have found their way to stackyards, barnyards, and outhouses, and are doing vast damage there. Even in gardens they are destroying the roots of plants and flower bulbs. . . . To exterminate them seems beyond the province of hope. Burning the ground where the destruction is greatest does no good, says one, they fly to their holes and ere long again appear; heavy rain does not drown them; some people, remembering how they disappeared after a similar but not so serious a plague about fifteen years ago, believe that a fall of slushy snow would kill them [a 'black' frost would have more effect], but slushy snow does not suit sheep, and such a remedy for the plague would of necessity involve great loss of stock. In some places more than the ordinary number of cats are kept, and these are credited with doing good work on the farmers' side. It is remarked that owls and hawks have been increasing all over the infected region; one informant mentions that in his locality the latter are as plentiful as crows; and in such an emergency all are gladly welcomed. But all that is being done in these various ways does not tell in any appreciable degree on the myriad swarms."[1] The only Peeblesshire farm I hear of being much affected by the plague is Fruid, in the southern part of the county. Rooks, I am told, devour great numbers of the young.

Desiring to see a few examples from different elevations, I applied to Dr J. R. Hamilton, of Hawick, who very kindly procured me a dozen from that neighbourhood. They were captured at various altitudes, from about 600 feet to close on 1000 feet above sea-level, and belonged without exception to the present species, *Arvicola agrestis*. One was cream-coloured (with black eyes), and the rest gave me the impression of being a shade darker than specimens I have examined from other localities. In acknowledging receipt of a couple I sent to the British Museum, Mr Oldfield Thomas, while unable to say that they present any peculiar features, adds that there can be no doubt about the species. My children have appropriated half a dozen of them as pets, and I don't think I ever before saw a wild animal take so readily to confinement. They exhibit practically no fear, and will sit on the hand for any length of time, regaling themselves on apple parings, bread soaked in milk, etc. The tender shoots of grasses they are very fond of, using the fore feet singly after

[1] Consult also Report by Board of Agriculture, March 1892.

the manner of a hand to bring the stems to the mouth and hold them in position. They show no desire to harm each other when in life, but the body of a dead companion is soon attacked and devoured. Furnished with strong chisel-like teeth, they are capable of making their escape in a very short time from almost any kind of wooden box. When disturbed or hungry they make a half grumbling, half squeaking noise, very much the same as guinea-pigs do, only not so loud.

The dimensions of this animal vary considerably, and do not appear to me to be always accounted for by age and sex. The following are a few measurements taken by myself from specimens captured in the months of January, March, and November:—

	♂	♂	♂	♀	♀
Length of head and body,	3·9 in.	4·0 in.	4·5 in.	3·75 in.	3·5 in.
Length of head alone,	1·2 ,,	1·2 ,,	1·4 ,,	1·2 ,,	1·1 ,,
Length of tail,	1·2 ,,	1·15 ,,	1·3 ,,	1·05 ,,	1·0 ,,

BANK VOLE.

ARVICOLA GLAREOLUS (*Schrcb.*).

My recent investigations among our micro-mammals have convinced me that the Bank Vole is common all along the valley of the Forth, and in all likelihood the same may be said of the Tay and the Tweed. It appears, however, to be in a great measure confined to the fertile belts in the lower parts of the valleys, becoming much scarcer or altogether absent in the upland districts, exactly where the Field Vole becomes most abundant. In the immediate neighbourhood of Edinburgh I find the Bank Vole the commoner of the two, and I am inclined to think this has long been the case, but there is no evidence to point to, as the earlier writers seldom distinguished between the two species—indeed, MacGillivray is the only one who does so with regard to the Forth area, and the only locality he mentions is near Bathgate, in the county of Linlithgow, where specimens were procured by Mr Durham Weir ("British Quadrupeds," 1838, p. 272). The only other Scotch locality given by MacGillivray for the animal is near Kelso, and on 6th May 1840 Dr Johnstone announced its occurrence at Mayfield in Berwickshire ("Proceedings" of the Berwickshire Naturalists' Club, vol. i, p. 214). Faldonside is another Border locality, in which, according to Dr Hardy (*op. cit.*, x., 278), it was abundant in 1883. During the last four years I have observed it at Rosetta and other places near Peebles, and Mr John Thomson has sent me one from Stobo, a few miles higher up the Tweed, where he tells me it is common about potatoe-pits during winter. Mr Harvie-Brown has sent it from Stirlingshire.

Seeing so little has been recorded of the Bank Vole in the neighbourhood of Edinburgh, the following facts from my own experience may not be without interest. In January 1886 I obtained one which had been killed by a weasel in Dalmeny Park, close to the Cramond ferry, and I then learned from the ferryman that the animal was common in the park, and did considerable damage during winter and spring to carnations and other flowering plants in his garden. The same complaint is made against it by Mr Bruce, gardener, Colinton House, from whom I have received many examples, and Mr Mackenzie, factor, Mortonhall, has also found it very troublesome in his garden of late. From Cramond I have obtained some very typical specimens, one of which I exhibited at a meeting of the Royal Physical Society on 15th January 1890. Since then I have trapped numbers in the following localities, namely, by the banks of the Braid burn below Comiston, on the Braid hills, by the roadside between Fairmilehead and Kaimes, at Dreghorn, at Lothianburn, and on the south side of the Pentlands, both by the roadside beyond Hillend and in the young plantation on the hill-side at Boghall. I have also obtained it at Gosford, in East Lothian. A bank on the sunny side of a wall is a favourite habitat, especially if well clothed with tussocks of cock's-foot grass (*Dactylis glomerata*). They may be seen sitting near the entrances to their burrows at all hours of the day, but the afternoon seems to be the time of their greatest activity. On a winter's day, if the sun has been bright, I can always depend on seeing numbers towards sunset feeding by the roadside which skirts the southern confines of Mortonhall grounds. As the spring advances they may be observed climbing the briars, thorns, and sapling elms, and nipping off the expanding leaf-buds.

The following are a few measurements taken from examples captured in January and March:—

Length of head and body,	3·2 in.	3·25 in.	3·25 in.	3·2 in.	3·5 in.
Length of head alone,	1·0 ,,	...	...	1·1 ,,	1·1 ,,
Length of tail,	1·45 ,,	1·5 in.	1·45 in.	1·6 ,,	1·4 ,,

While these pages were passing through the press (March 1892) I sent to the British Museum a few small Mammals, including a couple of critical specimens of the Bank Vole and a Lesser Shrew. Regarding the former, Mr Oldfield Thomas writes me as follows:—"They are very interesting specimens, and I was quite doubtful as to whether they were *Agrestis* or *Glareolus*, as they are so much less rufous in tint than the latter usually is. The teeth, however, show that they certainly are *Glareolus*. . . . Their tails are also a little shorter than usual." These peculiarities are characteristic, more or less, of a large proportion of the examples I have examined.

BROWN RAT.

Mus decumanus *Pall.*

The Brown Rat is only too well known wherever human habitations and industries have been established, finding a congenial home alike in town and country. It seems to be living more in the open fields now than formerly, and at times it increases to such an extent in certain localities as to become a serious agricultural pest, as has recently happened in East Lothian and the adjacent parts of Midlothian, where meetings of the farmers have been held to discuss the situation, and if possible devise a remedy (see numerous communications in the "Scotsman" during December and January 1890-91).

The first appearance of the Brown Rat among us does not seem to have been placed on record, but we may safely assume that the ports of the Firth of Forth were among the earliest localities in which the immigrants obtained a footing in Scotland; and we shall probably not be far wrong in referring the event to about the middle of the eighteenth century. By the beginning of the present century it was apparently only too common almost everywhere.

Walker, writing probably between 1764 and 1774, says of it, "First brought, as is reported, into Scotland in ships from Norway. Wherever it set up its abode, it entirely put to flight the *Mus rattus.*"[1] The following interesting account of its progress from Selkirk to the upper valley of the Tweed, as narrated in the "New Statistical Account" of the parish of Newlands (Peeblesshire, 1834, p. 137), is worth repeating "Zoology:—Under this head may be noticed the brown, or

[1] "Primum delatus, ut fertur, in Scotia, navibus e Norvegia. Ubicunque sedes suas figit, Murem *Rattum* penitus fugat" (Mammalia Scotica, p. 498).

Russian, or Norwegian rat, which a good many years ago invaded Tweeddale, to the total extermination of the former black rat inhabitants. Their first appearance was in the minister's glebe at Selkirk, about the year 1776 or 1777, where they were found burrowing in the earth, a propensity which occasioned considerable alarm, lest they should undermine houses. They seemed to follow the courses of waters and rivulets, and, passing from Selkirk, they were next heard of in the mill of Traquair; from thence, following up the Tweed, they appeared in the mills of Peebles; then entering by Lyne Water, they arrived at Flemington-mill, in this parish; and coming up the Lyne they reached this neighbourhood about the year 1791 or 1792." Neill includes it without remark in his Newhall list (1808).

BLACK RAT.

Mus rattus *L.*

Prior to the invasion of its haunts by *Mus decumanus*, the Black Rat infested all our towns and villages, and doubtless farm-steadings too. It seems to have been quite unable to live in competition with its more vigorous congener; and simultaneously with the rapid increase of the one, there took place a corresponding decrease of the other—cause and effect unquestionably—so that, by the early years of the present century, *Mus rattus* had practically ceased to exist in the coast towns, and a few years more sufficed to carry the extermination to its inland haunts as well. At the present time we have no proof of its existence on shore, though it is not improbable that a few now and again attempt to establish themselves in Leith and other ports, seeing they are known to exist in considerable numbers in vessels in the docks. A typical example (one of many) captured by a professional rat-catcher on board one of the Leith and Hamburg steamers while lying in Leith harbour in June 1890, was procured by Mr Eagle Clarke for the Edinburgh Museum, and recorded in the "Scottish Naturalist" (1891, p. 36); and I have seen another specimen, also taken on a Leith steamer, still more recently. Mr Thomas Hope, taxidermist, George Street, tells me that some nine or ten years ago, one, which had been captured in an Edinburgh skinnery, was brought to him for preservation. If his identification, which I have no reason to doubt, was correct, this is the last Edinburgh *Mus rattus* I have been able to trace.

The Black Rat was, of course, well known to Sibbald, Walker, and other early writers. Neill includes it in his Habbie's Howe and Tweeddale lists (1808 and 1815), and Stark (" Picture of Edinburgh," 1834) tells us that it " still inhabits the garrets of the high houses in the old city." Two years later Rhind dismisses it with the remark, " now rare " (" Excursions," p. 132); and in 1838 MacGillivray (" British Quadrupeds," p. 238) wrote thus—" In Edinburgh it appears to be completely extirpated, as I have not seen a specimen obtained there within these fifteen years."

In his list of Forfarshire animals (1813), Don says the Black Rat " is the only species I have seen in the town of Forfar, and it is not rare in all the inland parts of Angusshire" (Headrick's "Agriculture" of Forfar, App., p. 38).

The brown furred or tropical race, known as *Mus alexandrinus*, though abundant in the shipping in the Forth, apparently more so than the typical form, is not yet known to have obtained a footing on shore. The first record is that of an example received in Dec. 1888 by Mr Harvie-Brown from H.M.S. " Devastation," then stationed at Queensferry, and reported by Mr Eagle Clarke at a meeting of the Royal Physical Society on 19th March 1890; subsequently (August 1889) Mr Clarke had a cageful brought to him by a professional rat-catcher, who had just captured them on board one of the Leith and Aberdeen steamers ("Scot. Nat.," 1891, p. 36). I have since examined several others, also from Leith steamers. In Bell's "British Quadrupeds" the occurrence of this race in Britain is not positively asserted, though Lord Clermont, in 1859, had written—"Is often found in numbers in vessels from Egypt when discharging their cargoes of corn in British ports, but does not appear to spread in those towns, being probably kept down by the common species" ("Quadrupeds and Reptiles of Europe," p. 100).

HOUSE MOUSE.

Mus musculus *L.*

The House Mouse is only too common throughout the length and breadth of the district, establishing itself in and about human dwellings and other buildings, no matter how isolated they may stand. Having been for many years intimately connected with farming operations, I have often witnessed the havoc they commit in the stackyard, but their habits and economy are too well known to justify any remarks upon them here. Several specimens of a pale buff or cream-coloured variety were obtained for me in April 1890 by Mr R. S. Anderson of Peebles, from the farm of Lyne, where they were then in some abundance.

It is now impossible to trace the origin of this animal in the district. All that can be said is that its first appearance must have taken place many centuries ago.

WOOD OR LONG-TAILED FIELD MOUSE.

Mus sylvaticus *L.*

This timid but destructive creature is very common throughout most parts of the district, ranging from sea-level to a considerable elevation, and inhabiting woods, fields, and natural pastures alike. Though thus widely distributed, there can be no doubt it occurs in greatest numbers in the plains and warmer parts of the valleys, and practically avoids the damp upland tracts in which the Field Vole seems to delight. In the immediate neighbourhood of Edinburgh, where it is very abundant, I have recently trapped numbers among the furze bushes on the Braid and Blackford hills, among rough grass by the Braid burn, in the woods at Dreghorn, and at the foot of the Pentlands near Swanston; and have otherwise captured or identified it near Balerno, near Currie, at the head of Bonaly glen in the Pentlands, and in the woods at Rosslyn, Glencorse, Penicuik, etc. Many specimens have also been obtained for me in the garden and grounds at Colinton House, and in the woods and cottage gardens in Dalmeny Park. In East Lothian, where it is also abundant, I have trapped it on Luffness Links near Aberlady, and in the woods at Gosford; while in Fife I caught one at Otterston last August, and have lately detected it in the woods at Broomhall near Dunfermline, and in the neighbourhood of St Andrews. In Peeblesshire I have observed it at Macbiehill and at Eshielshope; and Mr J. Thomson tells me it is common at Stobo.

I have thrice had examples handed to me which were captured in dwelling-houses during winter, and have often seen its nests turned up by the plough. Having trapped them commonly in January and February during frost and snow, I conclude it does not hibernate in the true sense of the word, but we know it lays up stores of food for winter consumption. Being strictly nocturnal, these pretty little animals, though so abundant, are—unlike the Voles—seldom seen abroad in the daytime. During the last four months I have kept several in a cage with a covered-in portion at one end. In this den they have formed a nest of cotton and other soft materials, in which they pass the day snugly curled up and apparently fast asleep. After dark they come out to feed, and remain very active throughout the night, even although the gas be burning brightly in the room. When feeding, the motion of the under jaw is so rapid as almost to amount to vibration. Some weeks ago one escaped from the cage, and has since lived at large in the room, hiding itself during the day in a fold of the window-curtain. When surprised on the floor at night it climbs the curtains with astonishing rapidity, runs along the picture-rods, and, with a knowing look, sits up in kangaroo-fashion cleaning its face with its paws.

Among those that have passed through my hands I have noticed considerable variation in size and also in colour, some being much darker than others, the result of more black on the tips of the hairs. Probably these differences of tint are connected with the seasonal changes of fur.

Mus sylvaticus is included in Neill's list (1808), and in Rhind's list (1836).

HARVEST MOUSE.

Mus minutus *Pall.*

My efforts to obtain specimens of this interesting little animal from the district have proved singularly unsuccessful, and I find myself practically unable to add to the few records already in existence. Not only must it be very local, but I do not think it can be anywhere numerous, and it would seem to have been more easily procured in MacGillivray's day than now.

In Rhind's list of mammalia found in the immediate neighbourhood of Edinburgh ("Excursions," 1836, p. 132), "Mus messorius," the "Harvest Mouse," is entered with the remark, " not uncommon " against it; and MacGillivray states in his "British Quadrupeds" (1838, p. 257) that one was sent to him "from the neighbourhood of Edinburgh," and also that he once "found its nest in Fifeshire." In the "New Statistical Account" (Clackmannanshire, 1840, p. 9) it is included in a list of the animals of the parish of Alloa, and as pointed out by Mr Alston (Scottish Mammalia, p. 28), its size and weight correctly noted.

Mr Small, taxidermist, Edinburgh, assures me that about thirty years ago he received two, and within a week a third specimen for preservation. They were all from the same person, and Mr Small believes they were captured near Duns in Berwickshire. Curiously enough, I learn from Professor Duns that he once found a nest in the neighbourhood of the same town; this was prior to 1844. In August 1885 I found an unmistakable nest of this Mouse in a tuft of coarse grass growing under a hedge surrounding a corn-field behind Aberlady in East Lothian. It was about eighteen inches above the ground, and was supported entirely by the stems of the grass and a few of the twigs of the hedge.

Since the above was written, Mr D. F. Mackenzie, factor, Mortonhall, near Edinburgh, has informed me that in 1890 he observed a number of compact round nests among a heavy crop of oats on the home-farm there. They were placed one to two feet from the ground, and belonged to a small reddish mouse which he saw more than once sitting on the heads of the corn. Hoping they would reappear in the barley with which the field was last year cropped, a strict lookout for them was kept, but to no purpose, nor were they seen in any of the other fields on the farm. From Mr Mackenzie's minute description, I have no doubt the animals were a small colony of Harvest Mice, but it would have been more satisfactory had I been able to examine a specimen.

COMMON HARE.

Lepus timidus *L.*

The Common Hare is, and seems from time immemorial to have been, one of the best-known of our low-country animals. The volumes of the "Old Statistical Account" testify to its former abundance in the district, and no doubt the protection afforded by the game-laws, and the destruction of its natural enemies, tended to still further increase its numbers during the present century. A turn of the tide, however, has set in since the passing of the Ground Game Act in 1880, which gives the farmer the right to kill hares on the land he occupies. The result, which the proprietors are naturally enough deploring, has been a marked decrease in most localities, in some amounting almost to extinction. In the immediate neighbourhood of Edinburgh, fifteen years ago, I am certain I used to see twenty for every one observed at the present day. As a rule, it is now only where the grounds in the proprietor's own hands are of large extent that the Hare is to be seen in numbers. A close time, say from some date in February to a corresponding date in September, is urgently needed.

Though mainly an inhabitant of the plains, it occurs in the valleys of all our hill-ranges, extending in summer up the slopes of the hills themselves, even encroaching on the pastures of its congener, the Mountain Hare.

Coursing—the chasing of hares with greyhounds—is a favourite sport in the district. A pack of harriers also hunts the east of Fife, and there is at the present time a pack of beagles in Linlithgowshire.

Fleming tells us that in Scotland the skins were formerly "collected by itinerant dealers, and annually sold in the February market at Dumfries, sometimes to the amount of 30,000" ("British Animals," p. 21).

MOUNTAIN HARE.

Lepus variabilis *Pall.*

North of the Forth the Mountain Hare is abundant and indigenous among the Grampians, where I have seen it on many occasions, especially on the hills near Callander. Colquhoun, from what he says in his "Lecture on the Feræ Naturæ of the British Islands," would have us understand that in his young days it was very scarce on the Loch Lomond hills. In 1822 he "had shot over the whole rugged ground at the head of Loch Lomond without moving a single blue hare, barring the hermit on Ben Voirla's crags." It is included, however, in an excellent list of the animals of the parish of Luss, written nearly thirty years before the above date ("Old Statistical Account," xvii., p. 247). Farther east I observed one—still very white—in the third week of April 1891 on a high ridge of the Ochils above Tillicoultry, and learned from a shepherd that the species is fairly numerous on that range.

South of the Forth it is abundant on most of the higher

parts of the uplands from Lanarkshire through Peeblesshire to Selkirkshire, and extends along the Pentlands well into Midlothian. There are some now on the Moorfoots also, but I have not yet had any indications of its existence on the Lammermoors, though I have made a number of inquiries on the point. Mr Harvie-Brown informs me that it is not uncommon in the central range of the Stirlingshire hills.

It is generally understood that we owe their presence on the southern uplands entirely to the action of a few of the hill proprietors, by whom they have been introduced at different times within the last fifty to sixty years. Alston dates its existence in the south of Scotland from about 1860, but this is much too recent, as the following extracts show. In Chambers's "History of Peeblesshire," published in 1864, the following interesting passage occurs at page 525 :—" The Variable or Alpine Hare is now not unfrequent on the hills, but is known to have been introduced from the north by the late Mr Clason of Hallyards about seventeen or eighteen years ago. The first of the species in Peeblesshire were set free by Mr Clason on one of the highest hills in the parish of Manor. The species seems now to be fully established and naturalised over a very considerable district, extending many miles from the original spot." It would appear, however, to have been known in Manor a number of years before the date here assigned, as it is included in a list of the quadrupeds of the parish published in the "New Statistical Account" in 1834. An extract from one of Mr Alston's note-books, published in the "Proceedings" of the Natural History Society of Glasgow, vol. v., p. 73, records "that a Mr Hunter over at Glenbuck [on the borders of Lanarkshire and Ayrshire] turned out a number" about 1861. Mr B. N. Peach tells me that they increased very rapidly in that district, and that when living at Muirkirk, about twenty-five years ago, he found them quite plentiful. A few were also turned down by Mr Cowan about twenty-four or twenty-five years ago on the Silverburn hills, the highest of the Pentlands. From these three, and probably other points of introduction, the species has now spread over the greater part of the southern hill-country, where I have myself frequently observed them at various times of the year. In Peeblesshire I have recently come across them on the hills above Glen Sax and at the head of Manor; while in Selkirkshire I met with a few on Ettrick Pen and the hills above Tushielaw in June 1889. On the Pentlands they are well known as far east as the Cairn hills on the one hand, and Scaldlaw on the other; and Mr Cowan's keeper tells me they are still spreading. There are now a few on the north Black-hill, and on the south side of the range he saw one on Capelaw during the winter of 1889-90; another came under my own observation recently on a spur of Carnethy. On 1st January 1889, I made an excursion to the tops of Craigengar and the West Cairn-hill for the express purpose of seeing these Hares in their white coats, and was rewarded by the sight of several. Mr P. Adair, who has shot many of them on the latter hill during the last nine or ten years, informs me he has there seen a hybrid between this and the Common Hare, and in January 1891 I examined an undoubted example from near Cardrona in Peeblesshire.

RABBIT.

Lepus cuniculus *L.*

At the present time the Rabbit is perhaps the most ubiquitous of all our mammals, abounding alike on the islands of the Forth, and the dunes by the shores of the firths and estuaries; in the fields and woodlands of the plains; and among the rocks and pastures of the hills, where it lives at almost all elevations. From Sibbald's statement (quoted below) we may infer that it was also common and widely distributed in the district in the seventeenth century, though probably much less so than now; but I am inclined to think that between that time and the early part of the present century there was little if any increase in its numbers, except perhaps in a few localities. According to Don, it was rare in Forfarshire in 1813 (Headrick's "Agriculture" of the County, App., p. 38). A combination of circumstances, however, among which the destruction of its natural enemies has probably not been the least important, has since favoured its increase, and now it can only be kept within bounds by systematic trapping and snaring.

On 23rd May 1891 I found a Rabbit's nest at the foot of Auchinoon hill, in the parish of Midcalder, in an exceptional position. It was placed in the centre of a tuft of coarse grass, in what might have been a hare's "form," without the semblance of a burrow. In it were five young ones—blind and naked—enveloped in a mass of warm fur.

From Boece's "Description of Scotland," we learn that in the early part of the sixteenth century the islands of the Forth were "verie full of conies" (Holinshed's translation, 1805 edition, p. 13); and in Stuart's "Priory of the Isle of May," page xl, reference is made to a deed, by which in 1549 the prior of Pittenweem conveyed the island to Patrick Learmonth of Dairsy, in which deed the island is described as now waste, and spoiled by rabbits from which the principal revenue used to accrue, but of which the warrens were now completely destroyed and the place ruined by the English. Bones of the Rabbit found in a "kitchen midden" on Inchkeith ("Proceedings" of the Society of Antiquaries of Scotland, ix., 453), may point to it as an inhabitant of the islands of the Forth at a still earlier date, though they may merely have belonged to an animal that had made its burrow in the mound, and died there. In a charter granted on 10th November 1621 by James VI. in favour of the burgh of Peebles, we find "cunnings" and "cunningaries" specifically mentioned (Chambers's "Peeblesshire," p. 544). Sibbald, in his "Scotia Illustrata" (1684), says of the *Cuniculus,* "of these there is great plenty everywhere with us, especially on the coasts."[1] In the "Old Statistical Account of Scotland," the Rabbit is often mentioned, but chiefly as an inhabitant of maritime localities. In vol. xvii., p. 577, we are told that when Binning wood, at Tyninghame, was planted in 1707, "the East Links . . . . were a dead and barren sand, with scarcely any grass upon them, and of no use but as a rabbit-warren." The extensive sand-dunes stretching along the coast behind the village of Gullane, in East Lothian, have long been a noted warren. De Saussure, the Swiss naturalist, who visited these "grandes plaines de sable" in June 1807, in company with Patrick Neill, tells us that "un tres-grand nombre de lapins sauvages habitent ces dunes" ("Voyage en Écosse," vol. i., p. 162). Again, we read in Stark's "Picture of Edinburgh" (1834, p. 297), that the city market was then plentifully supplied with rabbits "brought chiefly

---

[1] "Horum magna ubique apud nos copia, in Littore præsertim."

from the extensive warrens at Gullane *Links* or downs in East Lothian."

Several varieties, doubtless the descendants of domestic animals run wild, are to be met with. One of these is thus referred to by Neill in the "Scots Magazine" for 1816, p. 170:— " On the Isle of May, in the entrance of the Firth of Forth, there exists a well-marked variety of the rabbit, distinguished not only by the great length of the hair, but by its silky fineness." Mr Agnew, for many years lighthouse-keeper on the island, tells me this form was still there when he left five years ago. I have recently observed a yellowish variety in some numbers on Gullane hill, and others with black fur near Cramond and on the Pentlands.

A hundred years ago the skin was the most valuable part of the animal; "The skins may be valued at 6s. a dozen, and the body sells at the rate of 5d. per pair" ("Old Statistical Account," parish of Dowally, vol. xx., p. 472). Now the skins are worth about 2s. a dozen, and the bodies are sold at 2s. 6d. a pair. Within the last two or three years the price of the skins has fallen by about one-half, owing to the large importation from Australia and New Zealand.

If, as seems highly probable, the Rabbit was originally introduced into Scotland, it was most likely by the monks. The monastery on the Isle of May was founded by David I. before the middle of the twelfth century.

Order UNGULATA.

RED DEER.

Cervus elaphus *L.*

But for the protection of the deer-forest, it is very doubtful if I should have been able to mention the Red Deer as still an indigenous animal anywhere in the district. Semi-domesticated animals are kept in a few of the parks of the nobility, but we must pass beyond Dunblane before there is even a chance of seeing the Stag on his native heath. The only deer-forest having any connection with the district is Glenartney, the southern portion of which touches the valley of the Forth, on the water-shed behind Doune and Callander. It has been fenced in about twenty years, and at the present time is said to contain fully 1000 deer. Stragglers are occasionally to be seen outside the precincts of the forest, but, as a rule, they do not wander far from it. I have myself observed them on the hills to the east of Loch Lubnaig, and Colonel Duthie informs me that he saw six, marching in line, on the braes of Doune, on the 22nd of July 1889—they were on the Doune side of the wire fence, which marks the march between Lord Moray's moor and the Glenartney forest. In the "Old Statistical Account" of the parish of Doune (xx., p. 49), it is recorded that: "On the sides of Uaighmor, the stag bounds along the heath;" and in Graham's "Sketches of Perthshire" (ed. 1812), it is stated to have been then (as now) occasionally seen in the neighbourhood of the Trossachs. "In hard winters," he says, "when provender is scarce, the Red Deer of the northern forests sometimes wander in quest of food and shelter, as far as Glenfinglas and the heights of Craig-vad" (see also the "Old Statistical Account" of the parish of Callander, 1794, vol. xi., p. 598).

About ten years ago Red Deer were introduced to the park at Hopetoun, Linlithgowshire, where I have seen them on several occasions. The keeper tells me there are twenty-six in the park at present, but that four years ago there were fully double that number. During the winter of 1889-90 a hind, doubtless an escape from Hopetoun, made its appearance in Dalmeny park, where it remained some months, but had ultimately to be shot owing to the damage it committed among the young trees. In the park at Dalkeith Palace, a single hind may now be seen feeding with the herd of Fallow Deer kept there.

In former times the Red Deer must have been abundant and generally distributed in the south of Scotland. Tradition tells us that during the Middle Ages the Scottish kings and nobles were wont to hunt deer in the immediate neighbourhood of Edinburgh, and doubtless such was the case, though there is little reliable historical evidence to point to. Such tales, for instance, as that of the royal hunt of Roslin, in which King Robert the Bruce is represented to have staked the forest and estate of Pentland against the head of Sir William St Clair, must be regarded as in the highest degree legendary (*vide* Wilson's "Annals of Penicuik," 1891, p. 165). The Red Deer, which was probably in most localities long survived by the Roe, must now have been extinct in the lowlands for many centuries. Even in the mountainous country around St Mary's Loch, it seems to have been practically extinct for at least two hundred years. Professor Walker, after informing us that, according to Bishop Leslie, numerous stags of great size were found in the Meggat district about the year 1578, adds that the last of that region, after wandering solitary among the mountains for about thirty years, and known to all the inhabitants, was killed on the neighbouring hills of Annandale in 1763 ("Mammalia Scotica," 1808). It must indeed have been rare if it existed at all in that district in the beginning of the eighteenth century, for in Dr Pennecuik's "History of Tweeddale," published in 1715, it is thus referred to,—" Upon the head of this water [Meggat] is to be seen, first, a house deservedly called Dead-for-cald; then Wintrop-burn; Meggit-knows; the Crammel, which seems to have been an old hunting-house of our kings, for I saw in the hall thereof a very large *Hart's-horn* upon the wall for a clock-pinn; the like whereof I observed in several other country men's houses in that desart and solitary place, where both *Hart* and *Hynd*, *Dae* and *Rae* have been so frequent and numerous of old, as witness the name of the hill, *Hartfield*" (ed. 1815, p. 248). Hartlaw, Hartside, and Hindsidehill are Lammermoor place-names (Muirhead's "Birds of Berwickshire," Introd., p. xv).

Remains of the Red Deer have been unearthed in almost every part of the district, thus proving what history and tradition vaguely indicate, namely, that the animal once roamed over the entire area. The following list of localities is taken from Woodward and Sherborn's "Catalogue of British Fossil Vertebrata"—Edinburgh, Elphinstone, Cockenzie, Drem, Athelstaneford, Seacliffe, Coldingham, Westruther, Kimmerghame, Whitrig Bog, Selkirk, Maxton, Linton, Uphall, Dundas Castle, Stirling, etc. Little more than a year ago I was shown several leg-bones, which had just been found on the Pentlands, a locality whence many examples of Red Deer remains have been procured—specimens from near Bavelaw, for instance, also came under my notice not long ago.

FALLOW DEER.

Cervus dama *L.*

Seeing the Fallow Deer is not an indigenous animal in the country, and exists only in a semi-domesticated state in parks specially enclosed for its reception, its right to a place in this memoir may be questioned. With Bell's "British Quadrupeds" as a precedent, the usual practice, however, has been to include it in local faunal lists, and I see no reason to depart from that rule in the present instance. After all, it is practically as much entitled to a place in our fauna as the pheasant, and its claims to that distinction are certainly quite as good as those of the Canada goose or the mute swan.

Without attempting to give a list of the deer-parks in the district, I may mention the following, with which I am personally familiar, namely :—the Duke of Buccleuch's park at Dalkeith, and the Earl of Morton's at Dalmahoy, both in Midlothian ; the Earl of Hopetoun's, at Hopetoun House, Linlithgowshire ; and Mrs Hamilton-Ogilvy's, at Biel, in East Lothian.

The regulation strength of the Dalkeith herd is 300, and at the present time it contains rather over than under that number. Their presence adds another to the many charms of that fine park, and I know few more enjoyable sights than to see them bounding through the tall brackens in the depths of the old oak-wood. Mr Chouler, the Duke's gamekeeper, tells me the bucks begin with great regularity to "bellow" on or about the 9th of October, and by the middle of the month they may be heard grunting in all directions. During still moonlight nights the park then resounds with their hoarse voices, the general effect being sufficiently wild, in my estimation, to afford genuine pleasure to the naturalist. The first fawns are almost invariably dropped on 16th June. The number of Fallow Deer in the Hopetoun park at present is only 140; fifteen years ago they numbered fully 250. In the Biel park there are between 200 and 300, and I understand the Dalmahoy park contains about the same number. These herds, which contain both spotted and uniformly dark animals, of course serve a useful as well as an ornamental purpose, and furnish their owners and the game-dealers with a constant supply of excellent venison.

In 1889 I observed Fallow Deer in Eshielshope, near Peebles, on the property of Sir John Hay, Bart. They were introduced, I am told, forty-two years ago, and at one time numbered nearly two hundred, but lately they have been killed down owing to their destroying young trees and the adjoining farm crops, and now only about a dozen remain.

So far as I am aware, the date of the introduction of the Fallow Deer into the district is not known. We have positive knowledge of it, however, as far back as 1283, for which year the accounts of the king's chamberlain record, among other expenses connected with the royal park at Stirling, an allowance for mowing and carrying hay and litter for the use of the Fallow Deer in winter (Cosmo Innes's "Scotland in the Middle Ages," p. 125). From an observation made by Walker in his "Mammalia Scotica," which is supposed to have been written between 1764 and 1774, it appears that Fallow Deer have been kept in Hopetoun park for at least a couple of centuries. The white and the black varieties, he tells us, had existed there for sixty years without intermingling, until the mottled form was introduced, from which time all three forms brought forth young differing in colour from their own kind. He also states that the dark variety was first introduced into Scotland by James VI. The Dalkeith deer-park is mentioned in the "Old Statistical Account" (vol. xii., p. 27).

ROE DEER.

CAPREOLUS CAPRÆA *Gray.*

At the present time the Roe Deer is locally not uncommon in the district. In Midlothian it is practically confined to the upper section of the country drained by the two branches of the Esk, the individuals now and again seen in other parts of the county being mere wanderers. From 1865 to 1872 I was very familiar with it on the wooded banks of the North Esk above Penicuik, where as many as eight or nine might occasionally be seen together. A few were shot annually, so that their numbers scarcely varied from year to year, but there is reason to believe a heavy toll has occasionally been levied from them during recent years. About three years ago, I startled one in the old haunts, and the head of another, which had been killed in the woods near Glencorse in December 1890, has since been shown to me. Wanderers may be seen almost every year crossing the Pentlands, and I have a record of one shot in Midcalder parish. On the South Esk it is well known in the country around Temple, and quite recently I had an excellent view of one in a large wood between that village and Gorebridge. It may also be seen from time to time in the adjacent parts of East Lothian (the woods at Humbie and Salton, for instance, are localities from which I have had it reported), but throughout the rest of that county it seems to be entirely absent, nor can I hear of it in the adjoining parts of Berwickshire, except as a rare straggler as far east, however, as the Pease dean woods (Letter from Dr Hardy). In Peeblesshire they have established themselves in most of the large fir plantations which now clothe the hill-sides on both banks of the Tweed, and a few are annually shot by the sportsmen of that district. In November 1888 I was delighted to see a party of six bounding through a thicket in the grounds at Portmore, near Eddleston, and a similar group may be seen in the woods at Dawyck. In Linlithgowshire, I am told, it is occasionally seen, chiefly in the more inland parts; and in most of the woodlands of Stirling and south-west Perth it is more or less common. It inhabits the extensive woods at Tulliallan, where I have myself seen it, and its appearance in the plantations of some of the adjoining properties is a common occurrence. In the west of Fife it is not uncommon in the neighbourhood of Saline, for instance; but in the east of the county it appears to be rare —a few, however, still exist in the woods at Falkland.

In olden times the Roe was, without doubt, much more abundant and generally distributed in the district than now, but the destruction of the forests and thickets, the growth of agriculture, and the loss of protection, gradually drove it from the southern section of Scotland, so that during the whole of the eighteenth century, and probably longer, it seems to have been entirely absent from our bounds, except in the mountainous country around Callander. In most localities the "Rae" probably long survived the Red Deer, but, apart from tradition and a few place-names, there is comparatively little evidence of its former abundance. I cannot recall any direct historical evidence for the area with which we are more immediately concerned, but as proving the existence of the animal in the south of Scotland during the reign of Alexander II. (1214-1249), I may refer to the oft-quoted agreement between the Avenels and the Monks of Melrose, by which the latter were expressly precluded from hunting Hart and Hind, Boar and Roe, in the forest of Eskdale (C. Innes's "Sketches of Early Scotch History," p. 103, and the Duke of Argyll's "Scotland as It Was and as It Is," 2nd ed., p. 52). Remains of the Roe seem to be less frequently brought to light than those of the Red Deer. The discovery by Dr Hardy of a portion of an antler in the vicinity of an ancient British camp at Oldcambus, in the extreme east of Berwickshire, is a fact of much interest ("Proc." Berw. Nat. Club, ix., p. 242). As already mentioned (p. 89), the animal is alluded to by Dr Pennecuik (1715) as a former inhabitant of Tweeddale, and in Chambers's "History of Peeblesshire" (1864, p. 525), we read, "Of the animals which have become extinct in Peeblesshire, tradition preserves the memory only of the Red Deer and the Roe. The latter seems to have survived after the extinction of the former. It is probably, however, at least two hundred years since the last really wild deer was killed in the county." The nearest parish in which I find it mentioned in the "Old Statistical Account" is Callander; "Roes," says the writer, "breed in our woods" (vol. xi., p. 598).[1]

Prior to the middle of the last century, comparatively few artificially-planted woods of any extent existed in the district. About that time, however, the planting of trees became very popular among the proprietors of the land, and in the course of the next twenty or thirty years thousands of acres in all parts of the country were utilised in this way. By the beginning of the present century many of these plantations were of sufficient growth to afford excellent shelter to such an animal as the Roe, which was now, so to speak, being invited to return to its former haunts. The return movement

[1] See also Graham's Sketches of Perthshire.

soon set in, and in the course of a few years the Roe had made its appearance in many localities from which it had long been absent.[1] In the "New Statistical Account" of the parish of Alloa (page 9), we read that "Roe-deer .... have been seen occasionally for more than thirty years in Tullibody woods," and the writer of the article on Tillicoultry, in the same volume (Clackmannanshire, p. 70), says of it, "occasionally seen in the neighbouring plantations." In the same publication it is included among the wild animals of Gargunnock and Fintry in Stirlingshire. "In Fife," writes Fleming (1828), "they have reappeared of late years, in consequence of the increase of plantations" ("British Animals," p. 26); and Professor Duns, in an article on the migration of mammals, contributed to "Science for All," mentions their subsequent periodical appearance in a plantation bordering on the banks of the Avon, in Linlithgowshire. From an incidental remark in Jackson's "Chivalry of Scotland in the Days of King Robert Bruce, including the Royal Hunt of Roslin," published in 1848, the date of its reappearance at Penicuik, on the south side of the Pentlands, may be fixed at from 1840 to 1845. "Deer in a wild state have," he says, "lately come to the woods of Sir George Clerk, Bart., about two miles from King Side Edge" (page 109).

[1] A return movement was noted before the close of last century in the valley of the Tay ("Old Statistical Account," Little Dunkeld, vi., p. 361).

Order CETACEA.

HUMP-BACKED WHALE.

MEGAPTERA BOOPS (*Fab.*) = M. LONGIMANA (*Rudolphi*).

THE true or "whale-bone" Whales mentioned in this memoir can only be looked upon as casual visitors to our waters,—wanderers from their proper habitats in the North Atlantic and Arctic Oceans. They appear to be all more or less migratory, but the North Sea scarcely falls within the area of their periodical movements. Except in a very few instances, the occurrences cited in the following pages have taken place during the autumn and winter months, September being the most productive. Semi-fossil remains of large Whales have been found on several occasions (see Milne Home's "Estuary of the Forth," p. 25).

With us the Hump-backed Whale is a casual visitant of very rare occurrence. Of the three examples that have been recognised in Scottish waters, two may be mentioned here, namely, one which was cast ashore about two miles north of Berwick-upon-Tweed on 19th September 1829, and the famous "Tay Whale," which for five or six weeks in the end of 1883 disported itself frequently in the Firth of Tay opposite Dundee, to the astonishment of the good folks of that town.

The Berwick specimen, which was described and figured by Dr George Johnston in the "Transactions" of the Natural History Society of Northumberland, Durham, and Newcastle-on-Tyne (vol. i., p. 6), was between 35 and 36 feet in length, 24 feet in girth, and had pectoral fins 9 feet long.

It was a female. In its stomach were six cormorants, and a seventh, on which it was presumed to have choked, was sticking in its throat. It was sold for £17, 2s. 6d., and yielded only about eighteen gallons of very inferior oil. In Bell's reference to this specimen ("British Quadrupeds," 2nd ed., p. 394) there are two mistakes, namely, that it was cast ashore near Newcastle, and was 26 feet long.

Notices of the "Tay Whale" appeared in most of the newspapers at the time, the best account perhaps being that in the "Weekly Scotsman" of 5th January 1884. Subsequently, in 1888 and 1889, a very elaborate account of it by Professor Struthers was published in the "Journal of Anatomy and Physiology" (vols. xxii. and xxiii.). It was a male 40 feet in length, with pectoral fins 12 feet long, and was believed to have been attracted to the Tay by the abundance of young herring then in the firth. Some idea of its great strength and endurance may be formed from the following facts:—After several fruitless attempts, the animal was at length successfully harpooned on 31st December (1883)—two, and finally three harpoons being shot into it. Large iron bolts, &c., were also fired into it, and hand-lances were driven three feet deep in its back. At first two six-oared rowing boats and a steam launch were made fast to it, and four or five hours afterwards a steam tug was added. With this heavy drag it swam wildly about the firth for a time, and then took out to sea, pulling all but the launch after it. For some time it pursued a northerly course till off Montrose, when it turned and proceeded towards the Bell Rock, then towards the mouth of the Firth of Forth, and finally turned north again when six or seven miles off the Carr Rock. One by one the harpoon lines had parted, and during the morning of 1st January, when a little to the south of the Bell Rock, the last gave way, and the Whale was again free, after being "fast" for nearly twenty-two hours to a dead weight of between twenty and thirty tons, which it was computed it had towed between forty and fifty miles. Of course it was wounded beyond the possibility of recovery. For the time being, however, it made its escape, and was not seen again for a week, when some fishermen observed the carcase floating off Bervie, and on 8th January towed it into Stonehaven harbour, where it was sold for £226 to Mr Woods, Dundee, who had it embalmed, and for the next seven months it was on exhibition in various towns in Scotland and England. The skeleton was presented by Mr Woods to the Dundee Museum. Seeing this specimen, during its endeavours to effect its escape, is known to have approached within a few miles of the Carr Rock, the species may be given a place in the Forth fauna.

SIBBALD'S RORQUAL.

BALÆNOPTERA SIBBALDI (*Gray*).

Sibbald's Rorqual, or the Blue Whale—the largest creature at present known to inhabit the globe—is another rare casual visitant to our shores, only three examples having been recorded during the present century. The large whale, 78 feet long, stranded at Abercorn, in the estuary of the Forth, in September 1692, and recorded by Sibbald ("Phalainologia nova," p. 33), in all probability belonged, as has been pointed out by Sir William Turner, to this species. Three undoubted examples, however, have since occurred in the Forth. The first is the huge animal, 80 feet in length, whose skeleton hangs in the Museum of Science and Art, Edinburgh. It was found floating dead at the mouth of the Firth in October 1831, and was towed ashore near North Berwick, and sold

to Dr and Mr Knox, by whom it was dissected (see
"Proc." Roy. Soc. Edin., 1833, vol. i., p. 14). Another, which
Professor Turner has identified from the nasal bones, preserved by Dr M'Bain, was stranded on the Fife coast at
Aberdour in July 1858 ("Report of British Association,"
1871, p. 144). And lastly, there comes the famous "Long-
niddry Whale," which was stranded a little to the west of
Gosford Bay in East Lothian, on 3rd November 1869.
During the fortnight it lay stretched on the beach thousands
of people flocked to see it, and doubtless many of my
readers, like myself, helped to swell the crowd. The carcase
was purchased from the Board of Trade for £120 by an
oil merchant in Kirkcaldy, who had it towed across the
Firth and flensed on the beach close to that town. Professor
Turner, who secured the skeleton for the Anatomical Museum
of the Edinburgh University, has given a very full description of the animal in the "Transactions" of the Royal
Society of Edinburgh (xxvi., pp. 197-251). It was a female,
measuring 78 feet 9 inches in length, and contained a
male fœtus 19 feet 6 inches long. Its girth was estimated
at 45 feet, and its weight at 74 tons. It yielded 16 tons
of oil.

COMMON RORQUAL.

BALÆNOPTERA MUSCULUS (*L.*).

In the Common Rorqual or Razorback we have another
rare straggler to the district, no specimen having been
identified, so far as I know, since 1848.[1] The earlier writers

[1] Van Beneden, in his "Histoire naturelle des Cétacés des mers d'Europe,"
1889, speaks of an example in the Firth of Forth in April 1880, but the
statement must, I fear, be one of the many inaccuracies which that work
unfortunately contains, as no such occurrence is known to Sir William
Turner, to whom I am indebted for valuable notes on this and allied species.

did not distinguish between this and the last species, and in the volume on "Whales" in the "Naturalists' Library," published in 1837, records clearly referable to each are brought together under the name of "Great Northern Rorqual." Of the examples there mentioned, the following are now generally referred to the present species, namely— one 46 feet long, stranded in the Firth of Forth a little to the west of Burntisland on 17th November 1690, and described by Sibbald ("Phalainologia," p. 29); another, "precisely of equal size," forced ashore very near to the same spot at Burntisland on 10th June 1761, and recorded by Neill ("Memoirs of Wernerian Society," i., 212) from a MS. account of it by Dr Walker; and a male, 43 feet long, stranded near Alloa, in the upper part of the estuary of the Forth, on 23rd October 1808, and described by Neill (*op. cit.*, i., 201). The only example since recorded seems to be the female, 54 feet long, which was cast ashore near Kinkell, about three miles east of St Andrews, on 8th January 1848, and described by the late Mr R. Walker ("Scottish Naturalist," vol. i., p. 107). In connection with this occurrence, it is worth noting that another whale, said to be of this species, went ashore near Aberdeen on 18th December 1847. The Razorback stranded "near Kingask, Fife, in 1848," of which Sir William Turner has some of the baleen (Alston, Scottish Mammalia, p. 17), and Walker's Kinkell animal, are understood to be one and the same.

In June 1752 a large whale was stranded near Eyemouth in Berwickshire, which was probably of this species (see Scoresby's "Arctic Regions," vol. i.); and Professor Turner informs me he has the skull of a specimen obtained at Bervie, Kincardineshire, in October 1889.

RUDOLPHI'S RORQUAL.

BALÆNOPTERA BOREALIS *Less.*

In September 1872 a whale, which Sir William Turner has since shown to have been an example of Rudolphi's Rorqual, was captured at Snab, Kinneil, about a mile from Bo'ness, on the Firth of Forth, by some men who, seeing it floundering in shallow water, proceeded to the spot, and, having fastened a rope round its tail, hauled it nearer the shore, and then killed it. The "Scotsman" of 26th September contained a notice of the occurrence. The length of the animal from the tip of the beak to the end of the tail was about 37 feet, and its girth about 15 feet. The carcase, after being stripped of the blubber, was secured by Professor Turner, who, in order to thoroughly clean the bones and free them from the oil they contained, had them buried in the Botanic Garden in a mixture of earth and leaves, in which they were allowed to lie till the summer of 1881. The skeleton was then prepared for the Anatomical Museum of the University, where it is now preserved. Although captured in 1872, it was not till the skeleton had been carefully examined ten years later that Professor Turner became satisfied "that the animal was the Cetacean named by zoologists *Balænoptera borealis* or *laticeps*"—see his paper read to the Royal Society of Edinburgh, 20th February 1882, and printed in the "Journal of Anatomy and Physiology," vol. xvi., p. 471, in which he minutely describes the specimen. This is the first properly authenticated example of the species taken on the British coasts, and is an addition to Alston's list of Scottish Mammalia.

LESSER RORQUAL.

BALÆNOPTERA ROSTRATA (*Fab.*).

The Lesser Rorqual seems to enter the North Sea more frequently than its congeners, and as a consequence more examples of it have occurred in our waters. It can only be looked upon, however, as an occasional visitant.

On 15th May 1832, one 14 feet in length was captured in the salmon stake-nets near Largo ("Magazine of Natural History," v., p. 570), and about two years later (February 1834) Dr Knox obtained a young one, 9 feet 11 inches in length, from "near the Queensferry" ("Proc." Roy. Soc. Edin., i., p. 63; and "Naturalists' Library," "Whales," p. 143). The next I have a note of was found in the sea, apparently dead, near the Bell Rock, on 7th September 1857, and taken to Leith; it was 14 feet 5 inches long—see "Proceedings" of the Royal Physical Society, i., p. 441, where it is described by the late Dr M'Bain. According to Alston (Scottish Mammalia, p. 18) another was caught in the Firth of Forth in 1858. On 8th September 1870, an example about 18 feet long, and of which the skull and baleen are preserved in the Anatomical Museum of the Edinburgh University, was stranded near Burntisland;[1] and in September of the following year (1871) one was taken at Dunbar (skull, etc., in Anatomical Museum); while in 1872 another was caught in the herring-nets off Anstruther (Alston, Scottish Mammalia, p. 18). In the Anatomical Museum there is also the skull of a young male from Elie in 1879. Still more recently one (27 feet long) which I had the satisfaction of seeing in the flesh, was stranded at Granton Quarry on 24th January 1888 ("Scotsman," 30th January), and in

[1] On 29th July 1869, one 13 feet long was stranded near Arbroath (Scottish Naturalist. i., p. 111).

November following a small example was obtained near Alloa; both, I understand, were secured by Sir William Turner.

In the autumn of 1874, when on the North Sea, not far from the mouth of the Forth, I observed a Whale rise to the surface several times to " blow." It was probably an example of this species.

BEAKED WHALE.

HYPEROÖDON ROSTRATUS (*Chemnitz*).

We now pass to the toothed Cetaceans, and the first species falling to be noticed is the Beaked or Bottle-nosed Whale, which appears to be an irregular but not very uncommon visitor to our shores in autumn. The "Proceedings" of the Royal Physical Society for 1885-86 (vol. ix., pp. 25-47) contains a valuable paper by Sir William Turner, F.R.S., on the occurrence of the species in the Scottish seas, in which he gives particulars of the following among other authenticated Scottish examples. It will be observed that, with one exception, they are females, each accompanied by a young calf.

1. An adult female, $28\frac{1}{2}$ feet long, accompanied by a young female 9 feet long, captured at Alloa on 29th October 1845, and identified by the late Professor John Goodsir (see paper by Wm. Thompson in the "Annals and Magazine of Natural History" for 1846, vol. xvii., p. 153, where it is mentioned under Lacépède's name, *H. Butzkopf*). As pointed out by Professor Turner, an erroneous date (1839), which originated with the late Dr J. E. Gray, has been very generally assigned to this specimen. Dr Gray (" Catalogue of Seals and Whales," 1866, p. 339) also referred to it as an example of his *Hyperoödon (Lagenocetus) latifrons*,

under which name it appears in the works of Bell and Alston; but Professor Turner, who, I understand, has the skeleton of the animal in the Anatomical Museum, states that the skull does not possess the broad lofty crests of Gray's supposed species, which is now known to be merely the adult male of *H. rostratus*. Bell, I observe, further states that the calf which accompanied this specimen was a male, whereas Thompson says distinctly it was a female. Neither does the skeleton of the mother appear to be in the Museum of Science and Art, as stated by Bell and Alston, but in the Anatomical Museum of the University.

2. A female, 26 feet long and 15 feet in girth, captured at Grangemouth on 23rd September 1879: examined by Professor Turner.

3. An animal said to be $14\frac{1}{2}$ feet long, found dead on the shore at Blackness on 24th September 1879, and supposed to be the young of the last mentioned.

4. Two examples, probably mother and calf, stranded at South Queensferry in September 1883: sold to an oil merchant in Kirkcaldy.

5. A young male, 20 feet 6 inches long (22 feet following the curvature of the back), found on the beach between Tyninghame links and Peffer burn, near Dunbar, on 4th November 1885 ("Scotsman," 5th November): procured by Professor Turner, and described in his paper above referred to.

In the "Scots Magazine" for 1808 (p. 37), the occurrence of an example of "*Delphinus bidens*" (Turton's name for the present species) was thus recorded by Patrick Neill:—"In the beginning of December [1807], during a strong breeze; a Bottlenose Whale (*Delphinus bidens*) twenty-one feet long, was stranded near Goulon Point, in East Lothian. The country people instantly stripped off the blubber, leaving the krang or carcase to those who should come after!"

SOWERBY'S WHALE.

MESOPLODON BIDENS (*Sowerby*).

Of this comparatively scarce species only one example is known to have reached the shores of the south-east of Scotland. It was found in Dalgety Bay, near Aberdour, on the north side of the Firth of Forth, in October 1888, by one of the Earl of Moray's gamekeepers. The head, skeleton, and viscera were procured by Sir William Turner, who gave a description of the specimen at a meeting of the Royal Physical Society in December following ("Proceedings," vol. x., p. 5), and subsequently described its stomach in the "Journal of Anatomy and Physiology" (vol. xxiii.). The animal was a male. Its extreme length in a straight line was 15 feet 1 inch, and its weight 15 cwts. The skeleton is preserved in the Anatomical Museum of the Edinburgh University.

As this Cetacean is probably a migratory species, visiting the shores of Northern Europe in the fall of the year, we may reasonably look forward to the occurrence of other examples in our waters at no very distant date.

BELUGA OR WHITE WHALE.

DELPHINAPTERUS LEUCAS (*Pall.*).

The Beluga can only be regarded as a casual visitant of extreme rarity, its claim to a place in the fauna of the district resting on the occurrence of a single (male) specimen in the Firth of Forth so far back as 1815. It was killed early in

June of that year in the upper part of the estuary, near Cambuskenneth, by salmon-fishers, who attacked it with firearms and spears. Hearing of the capture, Mr Robert Bald of Alloa promptly secured it, and had it forwarded to Professor Jameson of Edinburgh, and in December of the following year an account of it was communicated to the Wernerian Natural History Society by Dr Barclay and Mr Neill. Their paper, illustrated by two plates, on one of which is a sketch of the animal, was printed in the Society's "Memoirs" (vol. iii., pp. 371-395). For about three months it had been observed almost daily passing and repassing Alloa harbour, and it was often observed at Kincardine also. It generally passed up the estuary (in pursuit of salmon it was supposed) when the tide was flowing, and returned with the ebb. Measured in a straight line, its extreme length was 13 feet 4 inches. Its stuffed skin is still preserved in the Museum of Science and Art, Edinburgh.

The species being an inhabitant of high northern latitudes, and only a rare straggler to the European side of the North Atlantic, this specimen is likely still long to remain unique as a Forth example.

PORPOISE.

PHOCÆNA COMMUNIS *F. Cuv.*

The Porpoise is by far the best-known Cetacean we have, occurring abundantly in the estuaries or adjoining waters throughout the year; it is, indeed, the only species which can be regarded as common and resident. In pursuit of its prey it ascends both the Tay and the Forth, practically as high as the tide flows, and it is not unfrequently captured in the salmon stake-nets by the shore and in herring-nets at sea. To those who take advantage of the summer sailings on the Forth, the line of black fins appearing and disappearing in regular succession must be familiar. I have seldom gone an excursion of any extent, in any part of the estuary and firth from Alloa to the Isle of May, without observing a school of half-a-dozen or more rolling along in characteristic manner. In May 1887, while exploring the precipitous coast between St Abb's Head and Fast Castle in Berwickshire, I observed a couple of Porpoises fishing close in shore, and by remaining motionless for a few minutes had the pleasure of seeing them tumbling about in a pool within ten to twenty yards from the rock on which I stood.

The Porpoise was well known to Sibbald as an inhabitant of both firths, and he shows, from a charter granted by Malcolm IV. in favour of the monks of Dunfermline, that in those days the head of the animal was esteemed a great delicacy, and that it had also an economic value for the sake of the oil ("History of Fife and Kinross," 1803 ed., pp. 116 and 295).

KILLER or GRAMPUS.

ORCA GLADIATOR (*Lacép.*).

This species is probably a more frequent visitor to our waters than the few authentic records of its occurrence would lead us to suppose. Every now and then one hears of Grampuses being seen in the Firths, but owing to the vague way in which the name "Grampus" is used by the seafaring population of the district, these statements can scarcely be taken into account.

Sibbald, in his "Phalainologia nova" (p. 7), records the

occurrence of several "Orcæ" in the Forth (at Culross and Blackness) in May 1691, and from his description of the animals there can be no doubt they belonged to the present species (Van Beneden so regards them in his "Histoire naturelle des Cétacés des mers d'Europe," p. 441).

In the "Scots Magazine" for October 1814 (p. 733), Patrick Neill gave an interesting account of a herd of "Grampuses" which appeared in the estuary of the Forth in the beginning of that month. On the 6th, fifteen of them were killed at the mouth of the Devon, about two miles above Alloa, and of those which then escaped two were captured near Tullibody and two near Stirling. They were of various lengths, from 9 to 20 feet, and of both sexes. From the detailed measurements given of one of the largest, we learn that the length of the dorsal fin was 3 feet 3 inches, the length of the flippers 3 feet, and their breadth 2 feet 3 inches. From these facts, and the statements that "the back and sides were jet black, and shining; the belly pure white; and there was a large oblong white compartment behind each eye;" also that "there were two beautiful rows of teeth, 24 in each jaw, making 48 in all," there can be no doubt the animals belonged, as Neill inferred, to the present species, and not to the next, the Caaing Whale, as the writer of the "New Statistical Account" of Alloa (Clackmannanshire, 1840, p. 9) seems to have thought.

Fleming, in his "British Animals" (1828, p. 84), states that in the Firth of Tay, the Grampus "goes nearly as far up as the salt-water reaches, almost every tide at flood, during the months of July and August, in pursuit of salmon, of which it devours immense numbers." In all likelihood the animals on whose movements this statement was based, represented other species besides the present. The latest authenticated capture of the Killer in our waters of which I have a note is that recorded by the late Mr John Gibson in the "Proceedings" of the Royal Physical Society (vol. iv., p. 99). The record refers to a male having the following dimensions:—Total length along the curve of the back, 21 feet 10 inches; girth of body, 13 feet; height of dorsal fin, 3 feet 10 inches. It was captured about a mile west of Granton, on 18th March 1876, and "on being dragged ashore, while still alive, it gave forth shrill piercing cries, somewhat resembling in their sharpness a woman's voice."

A few years ago, I observed in the seaward portion of the Firth of Forth several Cetaceans, which, from the height and shape of their dorsal fins, I took to be of the present species.

PILOT or CAAING WHALE.

GLOBICEPHALUS MELAS (*Traill*).

The Pilot Whale may be regarded as an irregular spring and autumn visitant, though comparatively few authenticated instances of its occurrence have been recorded. There can be no doubt it is constantly confounded with the last species by the uninitiated, under the name of "Grampus."

The twenty-five Cetaceans mentioned by Sibbald as stranded at Cramond Island, in the Firth of Forth, in 1690 ("Phalainologia," p. 10), are referred by Professor Van Beneden to this species (see his recent "Histoire naturelle des Cétacés des mers d'Europe," p. 508). The writer of the "New Statistical Account" of the parish of Alloa considered that the "school" of small whales which occurred in the upper part of the estuary in October 1814 were referable to the present species, but, as I have already shown (p. 108), Neill's description of them, published in the "Scots Magazine" at the time, makes it perfectly clear that they belonged to the last species. From a statement in Don's list of Forfarshire animals (Headrick's "Agriculture" of Angus, App., p. 39), it would appear that true "Ca'ing" Whales were stranded up the Firth of Tay prior to 1813.

The "Zoologist" for 1856 (p. 5095) contains a description by Dr J. Hardy of a male *G. melas*, measuring 20 feet in length and 11 feet in greatest girth, which came ashore among the rocks of Greenheugh, a short way to the west of St Helen's church, Oldcambus, Berwickshire, on 29th March of that year. At the same time another—much smaller—also came ashore a few miles farther west in the vicinity of Thorntonloch, in East Lothian. In April 1867 a herd, supposed to consist of about two hundred animals, was observed in the Firth of Forth for about a fortnight. On the 19th the Volunteer Artillery at Portobello practised at them without success. The following day they were attacked by fishing crews and others from Prestonpans, Newhaven, and other villages, and no fewer than twenty-three of them slain, amidst scenes of intense and savage excitement. The bulk of the slaughter took place in the bay on the east side of Granton harbour. Three more were captured on the 22nd, and one or two others were cast dead on shore by the tide. These particulars are mainly taken from an account of the occurrence communicated by the late Mr E. R. Alston to the "Zoologist" (1867, p. 801). One of the animals (a female, 15 feet 2 inches in length), of which Mr Alston gives a description, was taken to Glasgow by a party of Newhaven fishermen and exhibited as a "Grampus," shoals of which, they said, were often seen about the Bass Rock, but it was very rarely they entered the Firth. Several of the animals captured on this occasion were secured for scientific purposes, with the result that our knowledge of the organisation of the species was greatly increased (see, for instance, Sir William Turner's paper in the "Journal of Anatomy and Physiology," vol. ii., and Dr Murie's treatise in the "Transactions" of the Zoological Society of London, vol. viii.). The skeletons of two of the animals are preserved in Edinburgh—one in the Museum of Science and Art, the other in the Anatomical Museum of the University.

The "Proceedings" of the Berwickshire Naturalists' Club, vol. vii., p. 509, contains a record by Dr Hardy of an example 14 feet long, which came ashore in October 1875 at Burnmouth, near Berwick; and on 3rd August last (1891) two small whales, which—as reported in the "Scotsman"—were stranded at St Margaret's Hope, near North Queensferry, also belonged to this species, as I am informed by Mr James Simpson, assistant to Sir William Turner.

WHITE-BEAKED DOLPHIN.

LAGENORHYNCHUS ALBIROSTRIS *Gray*.

Up to the date of the publication of Alston's list of Scottish Mammalia, no authentic instance of the occurrence of the White-beaked Dolphin in the Scottish seas was known. Since then several have been taken on different parts of our coasts, both east and west. Although it has not, as yet, been identified in the waters of the Forth, the fact that it has been captured off the mouth of the Tay on the one hand, and off the Tweed on the other, renders it highly probable that a few occasionally visit the seaward portion of the Forth also, and we may safely predict that its authentication for that area is only a matter of time.

On 7th September 1880 a young male was captured near the Bell Rock, and presented to the Kelvingrove Museum, Glasgow. This individual, which measured 5 feet 8 inches, was fully described by Mr J. M. Campbell at a meeting of the Glasgow Natural History Society on 30th November 1880, and in the "Scottish Naturalist" for January 1881 (p. 1).

In July 1881, an example was caught off Berwick, and in August 1883 another specimen—a young female—was also taken off Berwick and secured for the Kelso Museum, where its stuffed skin is preserved. The skulls of these two animals were handed over by the late Mr Andrew Brotherston to Sir William Turner, to whose "Notes" on the species, published in the "Proceedings" of the Royal Physical Society for 1888-89 (vol. x., p. 14), I would refer those who desire further information regarding the occurrence of this Cetacean in Scottish waters. The only examples I have myself seen in the flesh were an adult female and a young male, which were taken together off Stonehaven, Kincardineshire, in July 1888, and placed on view in the shop of Mr Anderson, fishmonger, Edinburgh. Both were purchased for the Anatomical Museum of the University by Professor Turner, who has given a minute description of them in his paper above referred to. The mother measured 8 feet 6 inches in length, and the calf 3 feet 11 inches.

It will be noted that this species has been observed on our coasts only during the months of July, August, and September.

COMMON DOLPHIN.

DELPHINUS DELPHIS *L.*

The Dolphin, being more of a southern species than most of the other Cetaceans here mentioned, is probably only an occasional visitant to our shores.

The "Dolphin," as distinguished from the Porpoise, was specifically mentioned by Sibbald as occurring in the Firths of Forth and Tay in the seventeenth century, and their relative sizes were correctly indicated. His words are as follows:—

Of these [Delphinidæ] in both these firths there are two sorts. The bigger beareth the name of Dolphin, and our fishers call them Meer-swines. The lesser is called Phocæna, a Porpess" ("History of Fife and Kinross," new ed., 1803, p. 115). In his "Phalainologia nova" (p. 6) he also mentions the "*Delphinus*," as distinguished from the "*Orca*" and "*Phocæna*," and gives an excellent figure of it, so that there is reason to believe some at least of his Dolphins were the true one. Attention may also be drawn to the fact that Don includes the species in his list of Forfarshire animals (Headrick's "Agriculture" of Angus, App., p. 39).

In the Museum of Science and Art, Edinburgh, there is exhibited a stuffed specimen of the Dolphin, labelled "Firth of Forth," but I have not been able to learn more of its history. It is understood to have been preserved at least thirty to thirty-five years ago.

From these somewhat unsatisfactory records, we pass to the following recent and authentic occurrence of *Delphinus delphis* in the Firth of Forth. In February 1887, a boating party observed a school of six or eight small Cetaceans swimming about in pairs in a bay on the Dalmeny estate between South Queensferry and Hound's Point, and succeeded in shooting one, which proved to be a female of this species, measuring in a straight line 5 feet 5½ inches. It was procured by Sir William Turner for the museum of the University, and is fully described by him in the "Proceedings" of the Royal Physical Society (vol. ix., p. 346).

BOTTLE-NOSED DOLPHIN.

Tursiops tursio (*Fab.*).

The following museum specimens furnish the only records I can find of the occurrence of this species within our bounds, namely:—Two specimens—a stuffed skin and a skeleton, perhaps taken from the same animal—in the Edinburgh Museum of Science and Art, labelled "Firth of Forth"; the skeleton of another, also from the Forth, which, according to Bell and Alston, formed part of the University collection formerly kept in the Surgeons' Hall; and three skeletons and a skull, all likewise from the Firth of Forth, in the zoological department of the British Museum.

Mr Eagle Clarke informs me that an entry in "The University Museum Register" shows that the specimen stuffed in the Museum of Science and Art was cast ashore at Portobello in the "year 1833-1834"; and it would appear, from Flower's List of Cetacea in the British Museum (p. 27), that the skeletons and skull in the national collection were purchased in 1866. The skull figured in the supplement to Gray's Catalogue (p. 73) is one of these specimens.

The Bottle-nosed Dolphin, though apparently only an irregular visitant to our waters, is probably less rare than has generally been supposed, and I think I may venture to predict that the capture and identification of fresh examples is only a question of time.

LIST OF PUBLICATIONS CONSULTED.

BOECE, HECTOR. Description of Scotland—contained in Holinshed's "Scottish Chronicle," 1577, and mod. edit.

SIBBALD, Sir ROBT. "Scotia Illustrata, sive Prodromus historiæ naturalis," 1684.

――― "Phalainologia nova, sive observationes de rarioribus quibusdam Balænis in Scotiæ littus nuper ejectis," 1692.

――― "History of Fife and Kinross," 1710; and ed. 1803.

PENNECUIK, Dr ALEX. "Description of Tweeddale," 1715; also new ed., 1815, containing list of animals, evidently by Patrick Neill.

WALKER, Rev. JOHN. "Mammalia Scotica," probably written between 1764 and 1774—contained in vol. of "Essays on Natural History and Rural Economy," 1808.

PENNANT, THOS. Sketch of Caledonian Zoology—prefixed to "Lightfoot's Flora Scotica," 2nd ed., 1792.

――― "British Zoology," ed. 1812.

(Old) "Statistical Account of Scotland"—Sir John Sinclair's —1791-1799.

"Scots Magazine" for years 1807 to 1817—containing monthly Memoranda in Natural History, by Patrick Neill.

"Wernerian Society's Memoirs," 1808-1837.

Royal Society of Edinburgh, "Transactions" and "Proceedings" of, various years—containing papers by Dr Knox, Sir William Turner, and others.

NEILL, PATRICK. List of the Animals of Habbie's Howe (Newhall, Carlops), published in the 1808 edition of Allan Ramsay's "Gentle Shepherd."

GRAHAM, Rev. P. "Sketches descriptive of Picturesque Scenery on the Southern Confines of Perthshire," eds. 1810 and 1812.

DON, G. List of Forfarshire Animals, published in the Appendix to "Headrick's General View of the Agriculture of the County of Angus," 1813.

DE SAUSSURE, L. A. NECKER. "Voyage en Écosse," 1821.

FLEMING, Rev. JOHN. "History of British Animals," 1828.

"Annals and Magazine of Natural History" (originally "Magazine of Natural History"), 1828 et seq.

"Natural History Society of Northumberland, Durham, and Newcastle-on-Tyne," vol. i., 1829.

"Berwickshire Naturalists' Club, History of," 1831 et seq.

STARK'S "Picture of Edinburgh," 6th ed., 1834; contains articles (understood to have been contributed by P. Neill) on the Objects of Natural History in the immediate neighbourhood of Edinburgh, and on the city markets.

RHIND'S "Excursions illustrative of the Geology and Natural History of the Environs of Edinburgh," 2nd ed., 1836—contains list of Mammalia.

HAMILTON, ROBT. "Natural History of the Ordinary Cetacea or Whales," 1837—a vol. of the Naturalist's Library.

MACGILLIVRAY, WM. "History of British Quadrupeds," 1838
—one of the vols. of the Naturalist's Library.

"New Statistical Account of Scotland," 1834-1845.

"Zoologist," 1843 et seq.

OWEN, Sir RICHD. "British Fossil Mammals and Birds," 1846.

FYFE, W. W. "Summer Life on Land and Water at South Queensferry," 1851.

Royal Physical Society, "Proceedings" of, 1854 et seq.

CLERMONT, LORD. "Guide to the Quadrupeds and Reptiles of Europe," 1859.

INNES, COSMO. "Scotland in the Middle Ages," 1860; and "Sketches of Early Scotch History," 1861.

CHAMBERS, WM. "History of Peeblesshire," 1864.

"Journal of Anatomy and Physiology," 1867 et seq.—papers by Sir William Turner and Professor Struthers.

"British Association Reports," various years.

"Zoological Record," 1864 et seq.

Natural History Society of Glasgow, "Proceedings" of, 1868 et seq.

Society of Antiquaries of Scotland, "Proceedings" of, vol. viii. (1868-70) et seq.

GRAY, J. E. "Catalogue of Seals and Whales in the British Museum," 2nd ed., 1866; and Supplement, 1871.

STUART, JOHN. "Records of the Priory of the Isle of May," 1868.

"Scottish Naturalist," 1871 et seq.

MILNE-HOME, D. "The Estuary of the Forth and adjoining districts viewed geologically," 1871.

LIST OF PUBLICATIONS CONSULTED

COLQUHOUN, J. "Lecture on the Feræ Naturæ of the British Islands," 1873.

BELL, THOMAS. "History of British Quadrupeds, including the Cetacea," 1874.

Zoological Society of London, "Transactions" of, vol. viii., 1874.

ALSTON, E. R. Article on Mammalia, in "Notes on the Fauna and Flora of the West of Scotland," 1876.

—— "Fauna of Scotland—Mammalia," 1880.

DOBSON, G. E. "Catalogue of the Chiroptera in the Collection of the British Museum," 1878.

HARTING, J. E. "British Animals Extinct within Historic Times," 1880.

FLOWER, W. H. "List of the Specimens of Cetacea in the Zoological Department of the British Museum," 1885.

Edinburgh Geological Society, "Transactions" of, vol. for 1886—contains paper on Reindeer and other bones from the Pentland Hills, by James Simpson.

VAN BENEDEN, P. J. "Histoire naturelle des Cétacés des mers d'Europe," 1889.

WOODWARD and SHERBORN. "Catalogue of British Fossil Vertebrata," 1890.

FLOWER and LYDEKKER's "Introduction to the Study of Mammals," 1891.

POLLOCK's "Dictionary of the Forth," 1891—contains an article on the Mammalia by W. Eagle Clarke.

INDEX.

	PAGE
Arrangement followed,	15
Arvicola agrestis,	65
,, amphibius,	64
,, ater,	65
,, glareolus,	70
Badger,	42
Bait for Micro-Mammals,	16
Balænoptera borealis,	13, 101
,, laticeps,	101
,, musculus,	99
,, rostrata,	102
,, sibbaldi,	98
Bat, Common,	19
,, Daubenton's or Water	20
,, Long-eared,	17
,, Natterer's,	13, 22
,, Whiskered,	23
Bear,	11
Beaver,	11
Beluga,	105
Boar, Wild,	11
Brock,	42
Caaing Whale,	109
Cachelot,	12
Canis vulpes,	37
Capreolus capræa,	92
Carnivora,	13, 34
Cat, Wild,	34
Cattle, Wild White,	12
Cervus dama,	90
,, elaphus,	87

	PAGE
Cetacea,	13, 96
Chiroptera,	13, 17
Conies,	84
Crossopus fodiens,	29
Cyclone Mouse-trap,	16
Cystophora cristata,	59
Deer, Fallow,	90
,, Red,	87
,, Roe,	92
Delphinapterus leucas,	105
Delphinus bidens,	104
,, delphis,	13, 113
Dolphin, Bottle-nosed,	114
,, Common,	113
,, White-beaked,	111
Dormouse,	13, 63
Elk,	12
Erinaceus europæus,	25
Ermine,	53
Felis catus,	34
Fox,	37
Foumart,	49
Globicephalus melas,	109
Grampus,	107
,, griseus,	13
Halichœrus grypus,	55
Hare, Common,	80
,, Mountain,	81

INDEX

	PAGE
Hedgehog,	25
Hyperoödon rostratus,	103
,, Butzkopf,	103
,, latifrons,	103
Insectivora,	13, 25
Killer,	107
Lagenocetus latifrons,	103
Lagenorhynchus albirostris,	13, 111
Lepus cuniculus,	84
,, timidus,	80
,, variabilis,	81
Longniddry Whale,	99
Lutra vulgaris,	40
Mammoth,	11, 12
Marten, Pine,	47
Megaptera boops,	96
,, longimana,	96
Meles taxus,	42
Mesoplodon bidens,	105
Mole,	31
Monodon monoceros,	12
Mouse, Harvest,	78
,, House,	76
,, Long-tailed or Wood,	77
Mus alexandrinus,	75
,, decumanus,	73
,, minutus,	78
,, musculus,	76
,, rattus,	74
,, sylvaticus,	77
Muscardinus avellanarius,	13, 63
Mustela erminea,	53
,, martes,	47
,, putorius,	49
,, vulgaris,	52
Narwhal,	12
Nomenclature followed,	15
Orca gladiator,	107
Otter,	40
Ox, Great Long-horned,	12

	PAGE
Phoca grœlandica,	13, 56
,, vitulina,	57
Phocaena communis,	106
Physeter macrocephalus,	12
Pilot Whale,	109
Pipistrelle,	19
Plecotus auritus,	17
Polecat,	49
Porpoise,	106
Publications consulted,	117
Rabbit,	84
Rat, Black,	74
,, Brown or Norway,	73
Razorback,	99
Reindeer,	12
Rodentia,	13, 60
Rorqual, Common,	99
,, Lesser,	102
,, Rudolphi's,	101
,, Sibbald's,	98
Sciurus vulgaris,	60
Seal, Common,	57
,, Greenland or Harp,	13, 56
,, Grey,	55
,, Hooded or Bladder-nose,	59
Shrew, Common,	26
,, Lesser,	27, 72
,, Water,	29
Sorex minutus,	27, 72
,, pygmæus,	27, 72
,, remifer,	29, 30
,, vulgaris,	26
Squirrel,	60
Statistical Account (Old), Extracts from,	10, 11
Stoat,	53
Talpa europæa,	31
Tay Whale,	96
Trap for Micro-Mammals,	16
Tursiops tursio,	114
Ungulata,	13, 87
Urus,	12

	PAGE			PAGE
Vespertilio daubentoni,	20	Whale, Blue,		98
,, mystacinus,	13, 23	,, Caaing,		169
,, nattereri,	13, 22	,, Hump-backed,		96
Vesperugo pipistrellus,	19	,, Pilot,		109
Vole, Bank,	70	,, Sowerby's,		105
,, Field,	65	,, Sperm,		12
,, Water,	64	,, White,		105
,, ,, black variety,	65	White Cattle,		12
		Wild Cat,		31
Weasel,	52	Wolf,		11
Whale, Beaked or Bottle-nosed,	103			

Printed by M'FARLANE & ERSKINE, *Edinburgh.*

www.ingramcontent.com/pod-product-compliance
Lightning Source LLC
Chambersburg PA
CBHW020135170426
43199CB00010B/753